research in conservation

The Getty Conservation Institute

Effects of Light on Materials in Collections

Data on Photoflash and Related Sources

Terry T. Schaeffer

2001

Dinah Berland: *Editorial Project Manager*
Elizabeth Maggio: *Manuscript Editor*
Anita Keys: *Production Coordinator*
Hespenheide Design: *Designer*

Printed in the United States of America
10 9 8 7 6 5 4 3 2 1

Getty Publications
1200 Getty Center Drive
Suite 500
Los Angeles, California 90049-1682

www.getty.edu

The Getty Conservation Institute
The Getty Conservation Institute works internationally to advance conservation in the visual arts. The Institute serves the conservation community through scientific research, education and training, model field projects, and the dissemination of information. The Institute is a program of the J. Paul Getty Trust, an international cultural and philanthropic institution devoted to the visual arts and the humanities.

Research in Conservation
The Research in Conservation reference series presents the findings of research conducted by the Getty Conservation Institute and its individual and institutional research partners, as well as state-of-the-art reviews of conservation literature. Each volume covers a topic of current interest to conservators and conservation scientists. Other volumes in the Research in Conservation series include *Biodeterioration of Stone in Tropical Environments: An Overview* (Kumar and Kumar 1999); *Inert Gases in the Control of Museum Insect Pests* (Selwitz and Maekawa 1998); *Oxygen-Free Museum Cases* (Maekawa 1998); *Accelerated Aging: Photochemical and Thermal Aspects* (Feller 1994); *Airborne Particles in Museums* (Nazaroff, Ligoki, et al. 1993); and *Statistical Analysis in Art Conservation Research* (Reedy and Reedy 1988).

Library of Congress Cataloging-in-Publication Data
Schaeffer, Terry T. (Terry Trosper), 1939–
Effects of light on materials in collections : data on photoflash and related sources / Terry T. Schaeffer.
p. cm.—(Research in conservation)
Includes bibliographical references and index.
ISBN 0-89236-645-1
1. Art–Deterioration. 2. Artists' materials—Effect of light on. 3. Photochemistry. 4. Archival materials—Deterioration. 5. Museum conservation methods. I. Title. II. Research in conservation (Unnumbered)
N8558.5 .S33 2001
069'.53—dc21

2001001977

Contents

Foreword

Among the many environmental conservation measures taken to protect objects in collections, the control of lighting levels has probably received the most unanimous agreement among conservators and conservation scientists. The more sensitive certain types of objects are, the more agreement exists that they be subject to general lighting guidelines that have been confirmed by scientific research. Yet despite these widely adopted standards, the conservation literature contains few reports of experiments designed specifically to address the effects of photoflash and reprographic light on works of art and archival objects. Terry T. Schaeffer's thoughtful review of the literature fills that gap, allowing us to dispense with most lingering doubts on the subject. The calculations provided here on the duration, intensity, and frequency of photoflash and reprographic exposure allow us to conclude at long last that, for most materials at least, there are few deleterious long-term consequences. Any legitimate doubts that persist will have to be addressed in future research.

The twentieth century witnessed a flood of new materials for art making, many of which did not take permanence into account. Objects containing these materials are now raising important questions regarding lightfastness and photochemical instability, and the twenty-first century promises to accelerate that trend. For some contemporary objects, photoflash and reprographic reproduction cannot be dismissed as insignificant. This review thus provides an in-depth examination of the photochemical sensitivity of very wide range of products.

The Getty Conservation Institute originally commissioned this study for internal use. We were seeking a thorough literature review on lighting of modern and historical materials with an emphasis on photoflash to inform our own judgment in planning future materials research. When the review was completed, its potential value to the larger conservation community was recognized and a decision was made to publish it in the GCI's Research in Conservation series.

Conservators, curators, and collections managers are faced with constant challenges in establishing exhibition policies and in managing risk to sensitive materials. We hope this publication may assist in that endeavor by providing new data in some cases and making previously less-accessible information available in a single volume.

Timothy P. Whalen, Director
The Getty Conservation Institute

Preface

This publication was first inspired by a request for a proposal to perform a modest experimental investigation. The topic was to be the effects of photoflash light sources on materials in art and archival objects. The purpose was to identify highly sensitive materials and to determine the extent of flash exposure that might be considered "safe."

Concerns for possible deleterious consequences of light exposures due to flashlamps have been raised periodically in the conservation community. These concerns are frequently dismissed on the basis of simple calculations that use all-encompassing assumptions and anecdotal evidence, or by reference to fading data for materials that are not highly light sensitive. A brief search of the conservation literature indicated that numerical data existed for only a few types of materials that had been thoroughly investigated using well-defined flash exposures. However, the conservation literature, and especially the photochemical literature, contains a wealth of data on the effects of various light exposures on many substances found in art and archival collections. If the relevance of these data to exposures from photoflash and reprographic light sources could be established, then general guidelines regarding the efficacy of photoflash exposures might be deduced.

A very large bibliography was amassed in the ensuing literature search, which was completed in spring 1998. A few more recent publications were also included. The data in the references were evaluated with respect to their implications for the effect of flash exposure. A set of carefully defined assumptions was established to enable evaluations to be done for many of the materials considered. Because little of the data reviewed here was obtained using flashlamps, or for materials and conditions normally found in museums, the results discussed must be considered *order-of-magnitude guidelines* only. In any specific situation, collections care professionals need to rely on their own expertise and judgment in applying the information presented.

The information obtained in the course of interpreting the literature for the effects of photoflash proved to be more useful than anticipated; it became apparent that these data could also be used a broad-based resource on light exposure of works of art and archival objects. Thus the entire bibliography and a variety of relevant background information on the effects of light on the materials considered were integrated here. The resulting publication is designed for museum personnel charged with devising general as well as specifically photographic light-exposure policies.

Acknowledgments

I wish to extend my gratitude to the Getty Conservation Institute for supporting the perception that a more extensive search of the conservation and photochemical literature was needed and for encouraging the preparation of this volume, which contains much more information than originally envisioned. It is hoped that this material will eventually be accessible to all museum and archives professionals who make light-exposure decisions for their collections.

The search could not have been completed without the support of the staff of the Getty Conservation Institute's Information Center. My thanks to Carol Cressler, Valerie Greathouse, Teresa Negrucci, Thomas Shreves, Shannon Walker, and Jackie Zak for their consistently cheerful and willing help while they were dealing with the logistics of moving and access problems of their own.

Many people suggested references or called my attention to particular materials for consideration, among them Margaret Holben Ellis, Narelle Jarry, Ann N'Gadi, Linda Sutherland, Shelley Svoboda, Cara Varnell, and Marianne Webb. Others provided information on and helpful discussion of light sources and photographic and scanner lighting systems: Adam Avila and Peter Brenner, Los Angeles County Museum of Art; Pat Braden, Minolta Corp.; Harry Campbell, Document Reproduction Services; Merri Egbert, General Electric Co.; David Elwell, Leica, Inc.; Franziska Frey, Image Permanence Institute; Rick Garcia, Océ-USA, Inc.; Dan Johnson, University of California, Berkeley, Library; Linda Levy, Bulbtronics; Pat MacFarland and Vince Marino, Olympus America, Inc.; Lou Meluso, J. Paul Getty Museum; Andrea Morris; Michael Pliskin, Nikon, Inc.; Ed Taylor, Norman Enterprises, Inc.; Steve Weintraub, Art Preservation Services; Don Wiedrich, Xerox Corp.; and the technical service staff members of Eastman Kodak Co. and Polaroid Corp.

Marco Leona of the Los Angeles County Museum of Art graciously translated a technical paper from Italian, reading aloud without advance preparation as I typed. Shin Maekawa, of the Getty Conservation Institute, provided translations of some sections of references in Japanese. Elizabeth Maggio contributed expert editorial advice. The staff at Getty Publications and its consultants provided the essential service of turning this technical manuscript into a book. Dinah Berland, editorial project manager, prepared the final manuscript with an experienced eye; Hespenheide Design shaped its presentation; and Anita Keys, production coordinator, shepherded the book into print.

As in all my conservation research endeavors, Victoria Blyth Hill provided moral support and a wealth of knowledge and experience; John Twilley contributed thoughtful comments; and Jim Druzik was always willing to engage in lively discussion of ideas and possibilities and to provide suggestions for presentation. Of course, any errors of citation and interpretation remain my own.

Terry T. Schaeffer

Introduction

As a form of energy, light has the potential to cause changes in objects that absorb it, including art and archival materials. Faced with this fact, museum professionals responsible for conserving collections occasionally suggest, facetiously, that the objects in their care would be better off kept in the dark in a rigorously controlled climate. In reality, conservators must deal with the fact that light-sensitive objects have finite display lifetimes and must seek to avoid unnecessary light exposure for such materials. This is part of the process of risk management, a combination of tasks designed to ensure the greatest possible viability and use of collections (Michalski 1994).

Up to now, no systematic effort has been made to review the photochemical literature and apply the information presented there to an assessment of the possible deleterious effects of exposure to light from photoflash and reprographic sources on art and archival materials; consideration of such exposure as a possible source of damage is an appropriate component of risk assessment in caring for collections. The purpose of this publication is to provide such information in a context that is useful to conservators and other museum professionals. To help the reader evaluate the data presented, a brief review of relevant photophysical and photochemical principles, of photometric and radiometric measurement, and of the spectral outputs of several light sources is provided. The approximations and assumptions used in the evaluation process are discussed in some detail, and examples of the different types of calculations are given.

To be managed effectively, risks must first be evaluated. This process always requires some degree of approximation and judgment based on relative magnitudes of different risks (Waller 1994). The risk of deleterious change to objects in museum collections due to absorption of ultraviolet light, visible light, or both is one of many risks that conservators must consider. For objects less sensitive to light, the risks associated with unnecessary light exposures are usually minor compared to other types of conservation challenges to preservation, and exposure to photoflash and reprographic light need not be avoided. For the most light-sensitive materials, however, a risk evaluation may be needed for each proposed light exposure. If the probable deleterious effects of the exposure outweigh the benefits for a particular object, that exposure should be avoided. Many art and archival materials will fall between the two extremes of light sensitivity; light exposure decisions for these intermediate cases may be based on rough approximations of extent of exposure and probable effects. The information assembled here offers conservators and collections managers some approaches to estimating potential damage to their collections from these types of light exposures so that informed risk-management decisions can be made.

Objects sensitive to light have a recognizable total display lifetime that is expressed in terms of the total light energy exposure that the materials can tolerate before their appearance or integrity is changed to an unacceptable degree. This unacceptable change depends on the nature of the object and its constituent materials. For colorants, and for materials that are white or colorless when new, the limit of minimum unacceptable change may be an alteration in color (change in hue or darkening or fading) that is barely detectable by the eye. Such a change might be quantitated, for example, by stating the maximum tolerable total color difference (ΔE) value and determining the total light exposure that causes a change of this magnitude. In other cases, a criterion for determining risk to an object due to light exposure may be the extent of change in another property, such as tensile strength. Useful background information on exhibition lighting and the damage it may cause has been presented by Cuttle (1988).

A relevant question for risk assessment is whether—or how much—exposure to incidental light sources will consume a significant, or unacceptable, portion of the total display lifetime of the object. One incidental source of light to which sensitive objects are likely to be exposed during their useful lifetimes is a flashlamp such as those used in photography or in reprographic systems. Thus the propriety of allowing unlimited, or undocumented, flash exposures in museum galleries is questioned periodically. As early as 1970, Hanlan responded to this concern experimentally when he tested the effect of tens of thousands of flashes of a xenon studio flashlamp on several pigments and on the International Standards Organization (ISO) Blue Wool standards used in the United Kingdom (see chap. 1, under "Blue Wool standards"). He compared his results to the effects of exposure to the equivalent total visible light energy from gallery lighting and to 150 hours of ultraviolet light from a mercury-arc lamp. He concluded that dyes and pigments that are generally considered stable would not be affected by normal use of photographic flashlamps, but that potentially fugitive materials "should be treated with great caution," suggesting that extensive photography of objects incorporating these sensitive materials be avoided where possible. Hanlan also noted that his results for the one fugitive pigment tested, for the most fugitive Blue Wool standards, and for his oil-plus-white pigment mixtures might have been due to multistep degradation processes. He suggested that these reactions may have been promoted by exposure to the electronic light flash and indicated that further investigation of these phenomena would be appropriate (Hanlan 1970).

Hanlan's study has been cited by many authors, including Thomson (1986) and Sancho-Arroyo and Rioux (1996). More recently, it was cited on the Internet (Conservation Distribution List 1996, response 10:18 to query 10:16) as support for the position that light exposure due to amateur flash photography would not normally be harmful to art objects on display and that gallery photography could be permitted in most cases. Oddy (1996) recently added the built-in light sources for handheld video cameras to the list of photographic lighting that could be allowed. In presenting his position for permitting gallery photography, Thomson used the sensible provisos that ultraviolet light should be excluded, maximum exposures should be specified, and care must be taken to prevent inadvertent exposure of other displayed objects that contain more-fugitive materials. Also, Sancho-Arroyo and Rioux concluded from their investigation of the heating effects of photographic flashlamps that the flashes themselves would have a negligible effect.

Concern for possible deleterious effects caused by exposing displayed objects to photographic light continues, however, due in part to anecdotal reports of deterioration (e.g., Ioanovici 1975). A recent evaluation of the ultraviolet output of several light sources used by studio photographers has added to these concerns. Neevel (1994) reported that several flashes from some flashguns used in galleries could deliver as much ultraviolet light to sensitive objects as an entire day's exposure to a museum lighting level of 50 lux of display lighting.

Saunders (1995) readdressed the question of light damage experimentally, choosing pigments and lakes of organic colorants that were expected to be significantly more light sensitive than the inorganic pigments used by Hanlan. He compared the effects on the colorants of two different levels of gallery lighting and of photographic flash exposure, after each light source had delivered the same total visible light energy. Most of the samples did not show significant differences in response to the three lighting regimens. Saunders concluded that in general the reciprocity principle is obeyed and that light from flashlamps is no more deleterious than an equivalent dose delivered slowly by gallery lighting. There were exceptions, however. In two cases the flash exposures were somewhat more damaging, and in three others the longer-term, slower rate of exposure from the gallery lighting caused a slightly greater change in appearance. The instances in which exposure to photographic flash was more damaging to colorants in art or archival objects than exposure to the equivalent amount of energy from gallery lighting are considered further here.

Michalski, in responding to the query posted on the Conservation Distribution List (1996, response 10:18 to query 10:16), evaluated the additional light exposure that objects in a gallery would experience due to a given number of typical amateur camera flashes per day, using specific assumptions. The most important of these were that (1) the principle of reciprocity applies; (2) the ultraviolet component of the light from the flash is unimportant; and (3) the tolerable increase in visible-light exposure is 10% for more-sensitive materials displayed in lighting of 50 lux, or 100% for objects in normal gallery lighting of 150 lux. Based on these assumptions, Michalski concluded that thirty-six hundred flashes, or a hundred thousand flashes per day over the course of an exhibition, would be needed to cause sufficient additional light exposures to be of concern for these two types of materials. All of these assumptions are examined in chapter 1.

Other considerations—such as copyright, aesthetics, or traffic flow—frequently take precedence in determining whether gallery photography should be permitted. Many conservators remain concerned about the effects of light exposures due to photographic flash, however, as evidenced by the responses to the initial query posted on the Conservation Distribution List (1996, query 10:16). In view of this, and of the extremely wide range of materials that are on display in different types of museums with varying levels of visitor interest, it is appropriate to reexamine the issue of the extent and possible effects of such exposures on the combinations of materials in art and archival objects. The impact of other types of lighting exposures to which art and archival objects may be subjected, such as ultraviolet photography for condition assessment, photocopying, and illumination used in other reprographic processes, also needs to be considered.

This literature search was undertaken to gather both conservation-related and photochemical research results pertinent to the issues previously outlined. The search was extensive but by no means exhaustive. Selected data from the literature and evaluations of these data to estimate tolerable exposures to flashlamp sources are summarized in chapter 2 for a wide variety of substances. More technical details of the evaluations for some materials are presented in chapter 3. Although emphasis has been placed on traditional artists' materials, representative modern materials that are beginning to appear in museum collections have also been included.

To ensure that the results of this investigation are of practical use to the widest range of professionals in the conservation community, some aspects of the physics of light and light measurement are summarized in chapter 1. An overview of the interaction of light with matter and brief descriptions of some photochemical and subsequent dark reactions that may result are also included. This information is not intended to supplant the use of textbooks on these subjects. Rather, it is provided to help readers follow the calculations used in the presentation of data on light sources in chapter 1 and on the effects of light exposures in chapters 2 and 3. A brief discussion of these results is given in chapter 4, which also includes a review of the information and assumptions needed to perform the calculations. Topics for experimentation to fill gaps in the available information are also suggested. Ultimately, it is hoped that this information will encourage collections care professionals to evaluate the effects of incidental light exposures on the art and archival materials in their own collections. The data evaluations might also be relevant to the design of more general light exposure guidelines, such as those suggested by Colby (1992) or Derbyshire and Ashley-Smith (1999).

Chapter 1
Background Information

Photophysics

The nature of light

Light is a form of electromagnetic energy that can be envisioned either as waves or as discrete packets of energy (quanta) that can travel through a vacuum or, depending on the circumstances, through materials. The term *photon* describes a quantum of light energy. Even though the quantum concept of light is used, a wavelength is always assigned to the light. The amount of energy in each photon—or carried by a wave—is inversely proportional to this wavelength, λ. If the wavelength of the light is known, the energy, E, of one photon can be calculated using the expression

$$E = \frac{2.0 \times 10^{-16}}{\lambda} \tag{1.1}$$

where E is in SI units of joules (J) when the wavelength is in nanometers (nm). If the total energy of the light of a particular wavelength that strikes an object is known, the number of photons with that energy can be calculated as follows:

$$\text{no. of photons} = \frac{\text{total energy at wavelength } \lambda}{\text{energy of one photon, } E} \tag{1.2}$$

By substituting the expression given in equation 1.1 for E into the denominator, the final equation becomes

$$\text{no. of photons} = \frac{(\text{total energy at wavelength } \lambda) \times \lambda}{2.0 \times 10^{-16}} \tag{1.3}$$

Again, the energy is in joules, and the wavelength is in nanometers. If, instead of total energy, the physical quantity specified is power (the rate per second at which energy is emitted by or arrives at an object), this equation can be used to calculate the number of photons per second:

$$\text{no. of photons/second} = \frac{(\text{total power at wavelength } \lambda) \times \lambda}{2.0 \times 10^{-16}} \tag{1.4}$$

The SI unit of power is the watt (W), which is equivalent to joules per second. These concepts are presented in general physics texts and in basic texts on the interaction of light with matter.[1]

The wavelength of visible light is an indication of the color of light that the human eye perceives, as well as the amount of electromagnetic energy per photon. Visible light ranges in wavelength from approximately 400 nm (blue) to 700 nm (red). If the wavelength is slightly shorter than that of visible light, it is called near ultraviolet; this has more energy per photon than visible light but, conversely, fewer photons for a particular total energy. If the wavelength is somewhat longer than 700 nm, the light is called infrared; it has less energy per photon than visible light and more photons for a given total energy. The rate at which light is emitted by a light source (i.e., energy per unit time) is the radiant power of the source.[2] Discussions from the conservators' point of view are presented by Feller (1994) and Thomson (1986).

The interaction of light and materials

Only light energy absorbed by a material can cause a photochemical change in that material. But absorption of light does not guarantee that a chemical change will follow. To consider the absorption of light energy by substances, the quantum description of light as photons is appropriate. A molecule will absorb a photon if the energy of that photon is exactly or very nearly equal to the difference between two energy states of the molecule. When the photon is absorbed, the molecule is said to be in an *excited electronic state*. In this condition, several different things can happen: the molecule can reemit the energy as fluorescence; the energy can be dissipated as heat; the energy can be transferred to another molecule; the molecule can lose a particular fraction of the energy and change to a somewhat lower, longer-lived excited state that subsequently emits the remaining excess energy as phosphorescence or dissipates the remaining energy as heat; or a bond between atoms in the excited molecule can break. Only this last possibility results in a chemical change in the molecule.

Often the possibility of chemical change is not very likely compared to the other possible outcomes. In these circumstances, it is said that the *quantum yield* of the photochemical reaction will be low. The quantum yield of any possible outcome is equal to the number of occurrences of that particular event divided by the total number of photons absorbed. When light interacts with matter, a large number of photons and molecules is involved even when the mass of the substance is small because of the small size of individual molecules. For example, suppose a million molecules each absorb a photon, and a hundred of those molecules change chemically (the rest of the excited molecules dissipate the extra energy as heat, or fluoresce, etc.). Then the quantum yield of the photochemical reaction is $100 \div 1{,}000{,}000 = 0.0001$, or 0.01%.

In general, to determine whether light striking an art object will cause a significant change (i.e., damage) in the object, we must know (1) whether the light energy is absorbed, and (2) what the quantum yield is for subsequent photochemical reaction. The first of these essential pieces of information is determined by comparing the emission spectrum of the light source with the absorption spectrum of the object; this tells what, if any, energy can be absorbed. The second piece of information is deduced from the results of quantitative photochemical experiments. The basic concepts involved are discussed extensively in books on photochemistry.[3]

The measurement of light energy

Because only the light energy that is absorbed by an object can initiate a change in that object, we must know how much light of the appropriate energies (or wavelengths) is incident on the object. Two different sets of units for measuring light energy are used in conservation disciplines, and each has its appropriate applications. One is the radiometric set, consisting of purely objective physical measures; many of these units incorporate the term *radiant*. This set of units is what is needed to evaluate the possible effects of light on materials. Unfortunately for this purpose, photographic light sources are usually rated in the other set—photometric units—the terms of which include the word *luminant* or its root. These units are a perceptual measure of light. They were developed to indicate the effectiveness of light in enabling people to perceive objects and color. Photometric units take into account the sensitivity of the normal eye to light of different energies, and they are valid only for visible light; that is, light of wavelengths between 400 and 700 nm. Additional information about the light source is therefore required to determine, for example, how much ultraviolet light it emits.

The units used to describe various physical aspects of light in each of the two sets are summarized in table 1.1. The basic radiometric unit is the watt (W). The watt does not describe light energy per se but rather

Table 1.1
Units used to describe light as energy or illumination[a]

Physical quantity	Description	Radiometric units (physical measure)	Photometric units (perceptual measure)
Flux	total power (energy/second) emitted by a light source	watts (W)	lumens (lm)
Intensity	power/unit solid angle, (conical shape of the light beam)	W/steradian (steradian is the unit of angle in three dimensions)	lm/steradian = candela (also referred to as candlepower)
Emittance	power/unit area emitted by a source (also called flux density)	W/m^2	lm/m^2 = lux; lm/ft^2 = foot-candle
Irradiance	rate at which light energy is incident on a unit area of surface	irradiance (W/m^2)	illuminance (lux or foot-candle)
Exitance	power/unit area emitted and reflected from a surface	W/m^2	lux or foot-candle
Brightness	intensity/unit of projected area	radiance ($W/steradian/m^2$)	luminance ($candela/m^2$)

At any wavelength in the visible region, to go from the **physical** quantity to the **perceptual** quantity, multiply the physical quantity by the luminous efficiency[b] at that wavelength.

[a] See also Snell (1978).
[b] The plot of lumnious efficiency as a function of wavelength is called the luminous efficiency curve (see fig. 1.1).

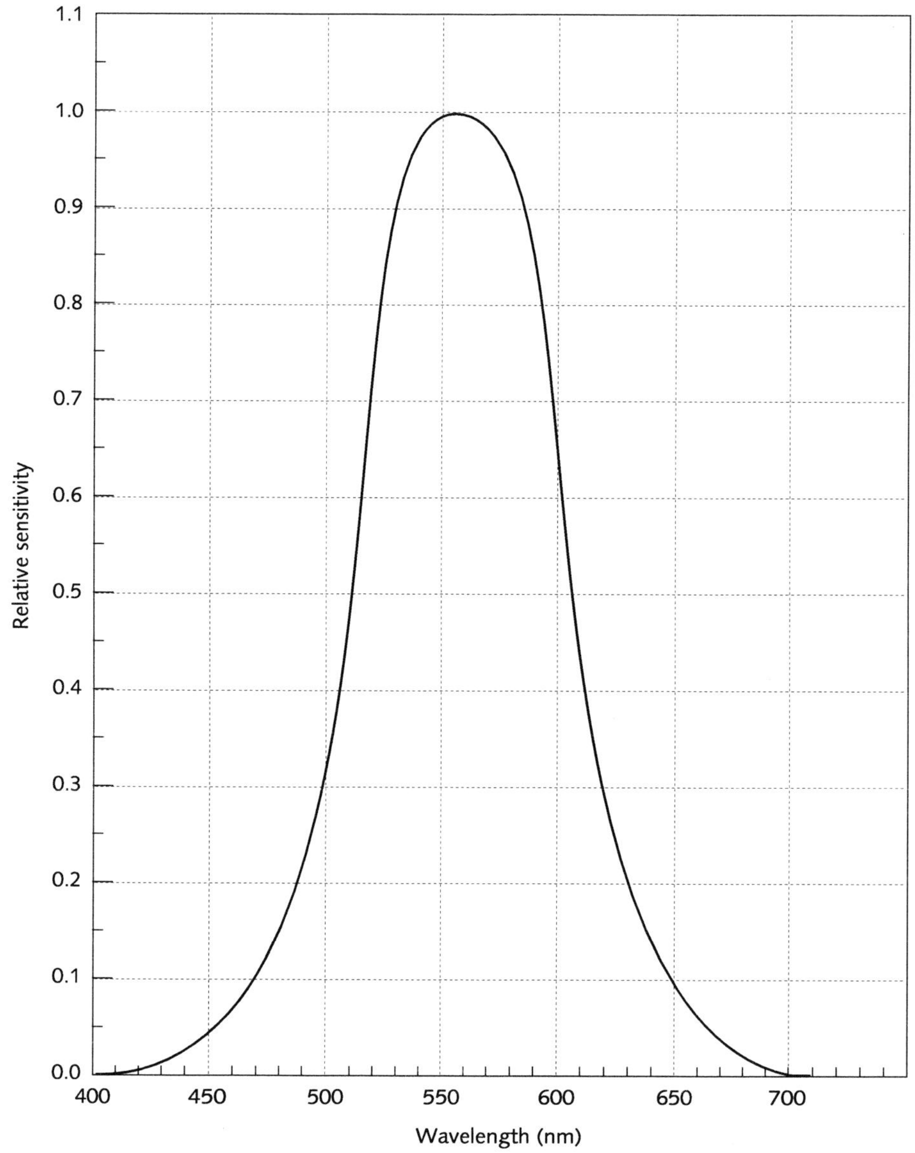

Figure 1.1
Photopic curve, showing the relative sensitivity of retinal cone cells to light, at daylight lighting levels, as a function of wavelength in the visible range (night vision is slightly different). The curve is a plot of the data at 10 nm intervals presented by Roberts (1997, table 3). The maximum has been interpolated.

power—that is, the rate at which light energy is emitted by the source or arrives at the object. The basic photometric unit is the lumen (lm). This is the total luminous flux, or the rate at which perceived visible light is emitted or reaches an object. It is only defined with reference to the *luminous efficiency curve*, also called the "photopic curve," shown in figure 1.1. This curve, which indicates how effectively the eye detects radiant energy at wavelengths throughout the visible range, has a maximum at 555 nm in the green-light region. A normal eye adapted to daylight can detect about 680 lumens per watt (680 lm/W) of radiant power at this wavelength. The luminous efficiency curve drops to 0 lm/W at 400 nm and at 700 nm. The dark-adapted eye responds similarly, but both the maximum wavelength and sensitivity are slightly lower. Thomson (1986) and Roberts (1997) present interesting discussions of both radiometric and photometric units. Roberts also includes a useful table of values for the luminous efficiency curve. Many of the units defined in table 1.1 are used throughout this book.[4]

Another way in which light sources are characterized is by specification of their color temperature. This classification is based on the physical concept of a *blackbody*. A perfectly black object absorbs all the electromagnetic energy that impinges on it. Depending on its temperature, it can also emit all wavelengths of radiation (not to be confused with the reflectance of light, which the hypothetical perfect blackbody does not do). Actually, any material that is heated will emit radiation in a range of energies. The higher the temperature, the higher the energy range, and therefore the shorter the wavelengths of emitted radiation. If the temperature of the object is high enough (thousands of degrees), the object will emit visible and ultraviolet light. The curve showing the amount of energy emitted per unit time as a function of wavelength, at a particular temperature, is known as a *blackbody curve*, and the temperature is the *color temperature*.[5]

Actual materials used as light sources may have, under appropriate conditions, emission curves very similar in shape to blackbody

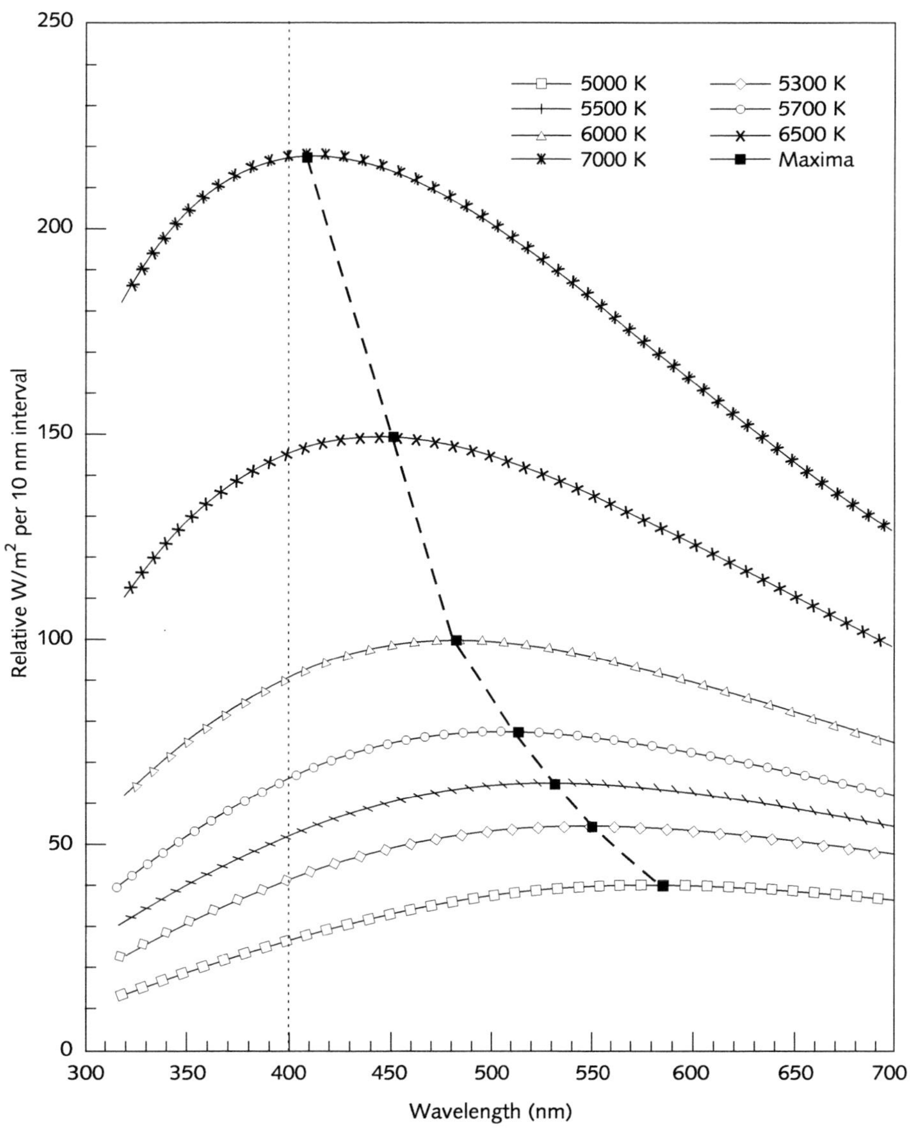

Figure 1.2
Blackbody emission curves for temperatures with curve maxima (indicated by solid squares connected by dashed line) in the visible region (indicated by the area to the right of the dotted line).

curves, some of which are shown in figure 1.2. The temperatures for the curves in this figure are in the range for sunlight, daylight, and xenon flashlamps. When a photographer uses a blue filter to lower the color temperature of a studio lamp, for example, from 7000 K to 5500 K, the corresponding blackbody curve is shifted to longer wavelengths.[6] The relative amount of blue and ultraviolet light is markedly reduced, and the total power emitted in the visible region is much lower. Also in this case, because the maximum value on the emission curve has been shifted closer to the wavelength of greatest sensitivity of the normal eye—and of typical daylight photographic film—the luminous emittance has not been decreased as much as suggested by the difference between the radiant emittance curves for the two temperatures. This sort of reasoning illustrates why photographic lighting is rated with photometric rather than radiometric units.

Additional information available about a light source may include the ultraviolet output, which is frequently given in mixed units of microwatts per lumen (μW/lm), that is, the rate at which ultraviolet energy is emitted per unit of perceived visible light. This might be provided as a spectral curve of power as a function of wavelength intervals in the near-ultraviolet region or as a single number for the entire near-ultraviolet wavelength range. In either case, to calculate the total ultraviolet power emitted, the total luminous flux of the source in lumens (see table 1.1) must be known; and to determine the amount of ultraviolet light at different wavelengths, the spectral curve is also required. Sometimes the entire spectral power output of the source (in microwatts per lumen), from ultraviolet to near infrared, is available plotted as a function of wavelength. Even in this case, the total luminous flux of the lamp must be known to be able to calculate the number of photons of light emitted at any wavelength. Only when this quantity is known can the effectiveness of a light source in causing a photochemical change in a particular material be determined accurately.

Photochemistry

Introductory surveys of photochemical processes may be found in organic chemistry textbooks at various levels (see, for example, Roberts, Stewart, and Caserio 1971; March 1985). Overviews from the conservation perspective have been written by Brill (1980) and Feller (1994, 177–85), and comments are included in various chapters in Mills and White (1994a, 1994b). Many texts on photochemistry and photobiology, such as Turro and Lamola (1977) and Wayne (1988), have basic introductory sections. The literature on the photochemistry of particular types of materials is vast but often not very easy to relate to conservators' concerns. Useful review articles and texts are cited in chapters 2 and 3 where appropriate.

A brief summary of the salient background is included here. As stated earlier, only light energy that is absorbed by a substance can lead to a photochemical change in molecules of that substance. The probability of the absorbed energy leading to a chemical change is, in most cases, much lower than the probabilities of other processes taking place, such as dissipation of the energy as heat or fluorescence. If a molecule absorbs light energy and has a relatively long-lived excited state—that is, if the absorbed energy is not entirely lost within about a microsecond—the chances of photochemical reactions occurring are greater. Frequently, the long-lived state is not the initial excited state but one of slightly lower energy, called a *triplet state*. These two types of states are distinguished by the orientations of the spins of the electrons involved in the bonds between atoms—a quantum mechanical concept that is not of concern here.

Occasionally, someone raises the question of whether *biphotonic processes* might occur in art objects during flash photography because of the higher intensity of flashlamps or strobe lamps compared to that of gallery light sources. For these processes to occur, a second photon must strike a particular molecule while it is still in the excited state from the first absorption (Wayne 1988, sec. 8.6). Biphotonic processes have been studied using powerful laser light sources that can be focused onto very small sample areas. Their occurrence is extremely unlikely during flash photography. A rough order of magnitude calculation will prove this point. Assume an excited state lasts, to be generous, for 10^{-5} second and that there are about 10^{18} molecules or monomers per unit area of the object being photographed. Also assume that 10^{15} photons of the appropriate energy strike that area per flash of 1/100 second duration. In this case, the largest fraction of molecules that could be excited per flash is $10^{15} \div 10^{18}$, or one in every one thousand molecules. This ratio is an overestimate because it does not take into account the fact that the absorbing molecules do not cover the entire surface of the object; there are spaces between them. The highest possible chance of a second photon hitting an excited molecule is $(1/1000) \times (1/1000) \times (10^{-5} \div 0.01) = 10^{-9}$, or one in one billion. The first two factors in the calculation are the probabilities of two photons hitting the same molecule. The third factor is the ratio of excited-state lifetime to the duration of the flash; this ratio is used because the second photon must arrive while the molecule is still in an excited state from having absorbed the first photon. This calculation helps to demonstrate that two-photon processes are of no concern unless the flash of light delivers roughly as many (or more) photons as there are molecules per unit area of object and does so in a very short time. Because the example overestimates the area covered by the molecules, even in situations of very high quantum fluxes from flashlamps, two-photon processes are unlikely.

The probability of a photochemical reaction following absorption of a photon of light is influenced, among other factors, by the surroundings of the excited molecule: whether it is in a liquid or a solid; if oxygen is accessible; if there are water molecules present; whether it is on a surface or at an interface; if it is near other molecules to which it can transfer its excitation energy or with which it can chemically react; and so on. The possibility—and nature—of photochemical reactions will also depend on properties of the excited molecules themselves: whether they contain particular groups of atoms that are likely to interact with one another because of their net charge or other factors; whether the molecules are flexible enough to allow these groups of atoms to get close enough to interact; what the distribution of electrons and net charge throughout the molecule may be; and so on.

Several categories of photochemical reactions are recognized. The main ones are cis–trans isomerization (switching positions of two of the four atoms or groups of atoms attached to the carbons sharing a double bond); rearrangements other than isomerizations of groups of atoms within the molecule; photocleavage into free radicals (resulting in each fragment of the molecule having a reactive electron that will cause further chemical reactions); photodecomposition into two smaller molecules without the formation of free radicals; photooxidation (formation of carbonyl or peroxide groups by addition of oxygen atoms); proton (H^+) abstraction (adding to the excited molecule a proton that has been removed from another molecule—effectively, a photoreduction of the first molecule); and photodimerization (joining of two like molecules, one of which has absorbed a photon). The excited molecule could also act as a photosensitizer, transferring the energy it absorbed to another molecule of a different material, which, in turn, would undergo a photochemical reaction (Spikes 1977). The type of photochemical reaction is determined largely, but not entirely, by the nature of the excited molecule and the presence of other substances required for the specific reaction. Examples of each category of photochemical reaction are given by Turro and Lamola (1977) and by March (1985). Discussion of the chemical mechanisms involved can be found in many texts, especially Wayne (1988).

The chemical changes listed here may (1) lead to the appearance of color in a material that is initially colorless, (2) cause a change in the color of a substance (e.g., photochromism), or (3) result in a decrease in the color of a material (i.e., fading). The reactions may not cause an immediate change in the appearance of an object, but the modified molecules may undergo additional chemical reactions that result in an appearance change over time. Also, in the case of photosensitization, the reactions of the photosensitized molecules could ultimately result in a change in the appearance of the object.

To predict the probable extent of change in an object due to exposure to photoflash or reprographic light, the quantum yield for a chemical reaction must be known or approximated. A reasonable value for the quantum yield (10^{-5}) has been assumed so the example calculations later in this chapter could be completed. Actual quantum yields for photochemical reactions of materials found in museum objects are seldom known. The quantum yield for a particular substance can depend on the physical state of the substance (solid, liquid, in solution, etc.) and on the presence of other chemicals required for the photochemical reaction to take place. If data are available on a particular material, they may be for a different condition of that material than the one being studied.

Determining quantum yields is a major activity in photochemical research. It requires accurate measurement of the number of photons absorbed by the reacting or sensitizing chemicals and of the number of product molecules formed. The evolution of instrumentation for this work has been remarkable during the past half century (Porter 1986). The application of this instrumentation to materials found in art objects (mostly mixtures of substances in the solid state), has tended to lag behind its application in other fields. For example, Egerton and Morgan (1970a) reviewed the fundamental principles of the photochemistry of dyes on the microsecond timescale, as revealed by flash photolysis of solutions. Flash photolysis studies on the nanosecond timescale were being performed in the late 1960s, but this technique was not applied to paper or dyed fabrics until after the development of diffuse reflectance spectrographic techniques in the 1980s (Wilkinson 1986).

In the many cases in which quantum yield data were not found for substances discussed here, two other approaches were taken. Order-of-magnitude values were employed, based on data published for the same material in a different physical state or for a similar material. In the absence of this type of information, which is often the case, data on the rate of fading caused by a particular light source over a relatively long time period were extrapolated to suggest a rough order-of-magnitude extent of change over the shorter time of exposure to photoflash or reprographic light. This approach incorporates the assumption that the reciprocity principle is applicable; thus, it will not produce valid estimates when reciprocity does not apply.

The physical basis for the reciprocity principle is that the determining factor responsible for change is the total amount of energy absorbed by the object rather than the rate at which the energy is absorbed. Saunders (1995) presents a good discussion of this principle, especially as it applies to flash photography versus gallery light exposure. Using this principle, the same extent of fading should occur, for example, if an object is illuminated at 5 lux for 100 hours or 50 lux for 10 hours.[7] Saunders and Kirby (1996) have demonstrated the validity of reciprocity for several light-sensitive colorants and for poor-quality paper. Certain types of color photographs appear to be a notable exception (Wilhelm 1993, chap. 2). Some individual examples of nonreciprocal behavior of particular materials are cited in chapter 2.

Light Sources

Flashlamp technology has changed significantly since the 1960s. A General Electric catalogue from that period (General Electric Company n.d.) contains many pages listing flashcubes and flashbulbs based on metallic foils or filaments, and only two pages showing electronic flashtubes, one of which lists pulsed xenon-arc lamps for professional studio use. They were linear or helical tubes with the smallest overall dimensions of at least a few inches. Built-in flash and miniaturized electronic xenon lamps were not available.

Today essentially all built-in flash units in single-lens-reflex and point-and-shoot cameras are xenon lamps. Studio photographers who photograph works of art frequently use xenon strobe lights to reduce heating effects. In his investigation of the characteristics of several electronic flashguns used by European commercial photographers, Neevel (1994) found that the range of ultraviolet output for the lamps he tested was 91–720 μW/lm; that is, none of these lamps would meet the conservators' rule of thumb of less than 75 μW of ultraviolet light energy per lumen of visible light for a light source used as gallery illumination. Because the flashguns emitted light for a small fraction of a second and had a wide range of illuminances, the total ultraviolet energy dose per square meter was in several cases quite low. Neevel presented these data in the useful format of the number of flashes needed to deliver a total amount of ultraviolet energy equivalent to that received during one hour of incandescent lamp illumination at 50 lux which is about 5 footcandles, (see table 1.1). The number of flashes ranged from nine to more than four hundred flashes.

As an example of the significance of this result, consider the worst-case lamp, which delivers an amount of ultraviolet light equivalent to one hour of gallery lighting in only nine flashes. If a photographer uses four of these lamps to illuminate the object, takes eight setup shots, and makes ten exposures, the object has experienced the equivalent of $[4 \times (8 + 10)] \div 9 = 72 \div 9 = 8$ hours of extra exposure to ultraviolet light; that is, its display lifetime has been reduced by about one day due to the photography. Whether this is an acceptable reduction will depend on the total display lifetime for the object. For a very sensitive object—for example, one with light sensitivity equivalent to that of Blue Wool no. 1 (a measure of a material's lightfastness, discussed later in this chapter)—the total display lifetime is roughly 4×10^5 lux hours, or 8000 hours at 50 lux. This equals one thousand 8-hour days, or about twenty exhibits of fifty days each. If the appearance of the object is to remain acceptable during exhibits over the course of a century, the object can go on display for a couple of months every five years during that century. If it is photographed with the worst example of today's flash technology at each exhibit, those exhibits would need to be one day shorter—or one exhibit should be half as long—to account for the photography, which uses up some of the display lifetime. Conservators might find this acceptable for objects that can safely be displayed at an illumination of 50 lux or greater. However, they may want to suggest restricting photoflash exposure for art and archival works that are immensely popular photography subjects or contain materials that are significantly more fugitive than Blue Wool no. 1. For example, four times the hypothesized sensitivity could shorten the display time by two exhibits. In consideration of future generations of museum visitors, curators need to be made aware of the ramifications of unlimited flash photography of highly light-sensitive objects.

When professional photographers use tungsten photoflood lamps, which emit a majority of radiation as heat energy, they generally minimize heating effects by using a variable voltage control to lower the lamp output between exposures. Thus, the total additional exposure to visible and near-ultraviolet radiation due to photography is negligible compared to that due to display lighting. For the purposes of this investigation, it has been assumed that tungsten filament lamp use in studio photography or professional gallery photography does not expose works of art to light of the intensity or wavelengths that would threaten the longevity of these objects. Gallery photography by professional groups using incandescent lamps should be monitored carefully to ensure that the practice of operating photoflood lamps at full power for the minimum amount of time necessary is always followed.

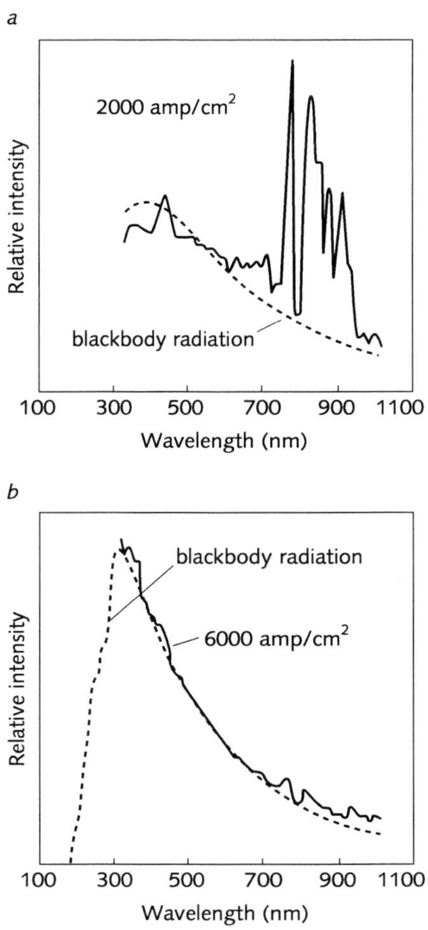

Figure 1.3
Emission spectra for two types of xenon-arc lamps: *a*, studio strobe light—UV-B and UV-C blocked by housing material; and *b*, flashlamp designed to have continuum emission similar to sunlight, for use in cameras. (Data on Heimann lamps supplied by Norman Enterprises, Inc.)

Xenon and tungsten lights are essentially the only sources used for photographing works of art; however, other light sources may be regularly used in other imaging activities involving art and archival objects. Xerographic photocopiers may have fluorescent, tungsten, tungsten–halogen, or xenon light sources. Scanners may operate with any of these sources or with light-emitting diodes. The spectral output of these latter devices can now be designed to resemble the spectral sensitivities of the three pigments responsible for human color vision; thus, they do not emit ultraviolet light and should not represent a significant threat to materials being scanned. The polymers used to fabricate these devices have not yet been designed to be sufficiently stable to ensure the required stability of spectral output over time.

The light sources in scanning and copying devices may illuminate the object for much longer times than flashlamps do during photography. Digital imaging processes may also require relatively long exposures of the objects to photoflood lighting as compared to the needs of studio photography (see chap. 2 for a further discussion of light exposure levels during reprographic procedures).

Mercury-arc lamps are used occasionally for examination of paintings during condition reporting and for ultraviolet photography to document earlier restorations. Carbon-arc lamps have been used to illuminate works of art in the past. Much of the research literature, especially the older literature (in the case of the carbon-arc lamp), describes the photochemistry of materials subjected to light from these sources.

Xenon lamps

The question of whether xenon flashlamps pose a threat when used in galleries or photography studios is examined here in numerical detail to illustrate the assumptions, reasoning, and calculations required to provide a context for decision making. The output of a xenon lamp depends on several physical parameters of the lamp and on its operation: the gas pressure, the lamp dimensions, the gap between the electrodes, the capacitance of the charging circuit (which influences the current in the discharge), and the voltage at which the lamp operates (Phillips 1983b). For lamps used in photography, these parameters are usually adjusted to result in a lamp output roughly equivalent, in the ultraviolet and visible regions, to a blackbody at 6000–7000 K (fig. 1.3). Variations in near-infrared output (where some strong emission bands are located) should not be of concern with these lamps, because the infrared energy is not great enough to cause photochemical changes, and the flashes are not of long enough duration to heat the object significantly.

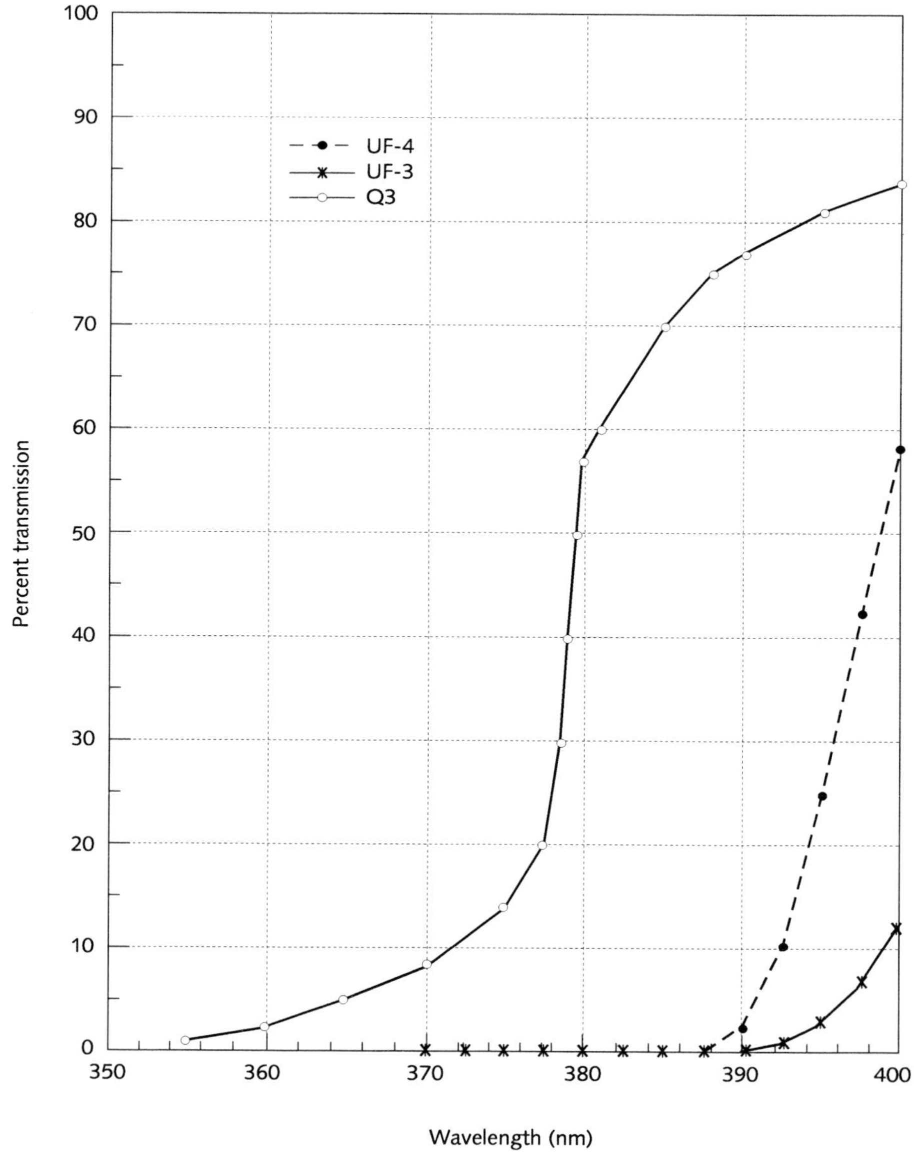

Figure 1.4
Transmission spectra of UV-blocking plastic filters (UF-3 and UF-4) and synthetic quartz lamp housing (Q3). (Plastic manufactured by Atohaas; spectra measured on a Beckman recording spectrophotometer; data on Q3 supplied by Norman Enterprises, Inc.)

The ultraviolet output of the lamp is reduced somewhat by the lamp housing. Figure 1.4 compares the percent transmission of two ultraviolet-blocking plastics commonly used for glazing—UF-3 and UF-4, manufactured by Atohaas—and a synthetic quartz xenon flashlamp housing called Q3. Although Q3 blocks much of the shorter-wavelength ultraviolet emitted by xenon, it transmits a significant portion of the longest ultraviolet wavelengths compared to the plastic glazing materials.[8]

Xenon lamp assemblies with window-glass housing or plastic exit windows that do not block near-ultraviolet light could be expected to emit near-UV light of wavelengths greater than 320 nm, as well as visible light. Because these near-UV wavelengths are capable of initiating photochemical changes in a variety of materials (see chap. 2), it is necessary to determine how much of this light may be emitted per flash.

To illustrate the calculation process, four different types of xenon lamps that may be used to illuminate art and archival materials for photography, research, or photoreproduction are considered here: a handheld camera, an Atlas Xenon Weather-Ometer CI 40000, a studio strobe with four lamps, and a typical 1980s-model photocopy machine. The conditions of exposure and the properties of the materials were chosen to represent possible real conditions and a variety of situations. Several numbers have been approximated and rounded off in these order-of-magnitude calculations. Assumptions have been made regarding, for example, the quantum yield for photochemical reactions, the distance of the object from the source, and the area of the object illuminated by the flash. Actual conditions that are very different from the assumptions could alter the numbers significantly, thus changing the conclusions reached (these considerations are discussed later in this chapter, with specific assumptions for particular materials identified in chap. 2).

The first case is a small, built-in xenon flash in a point-and-shoot instant camera. The flash is color corrected (that is, filters are part of the camera) to 5000 K, and the total beam candlepower is 700 lumens per flash. *Total beam candlepower* means that the lumens have been integrated over the entire conical beam of the directed flash; that is, the candlepower (stated in units of lumens per solid angle, or *candela;* see table 1.1) has been multiplied by the solid angle subtended by the beam. The camera is assumed to be about 1 m from the object (the point to which the automatic focus is set). In this case, an object with a surface area of 1 m^2 would receive all 700 lumens; the illuminance is therefore 700 lux (700 lm/m^2). Because the flash lasts only about 1/100 of a second, however, the total visible-light energy received from one flash is very low.

The ultraviolet energy content of the flash must now be considered. This value needs to be determined indirectly. The blackbody curve for 5000 K shows that the radiant power output between 320 and 400 nm is about 15% of the output in the visible range (between 400 and 700 nm); but the flash specification supplied with the camera is given in lumens—that is, in photometric rather than radiometric units. To deal with this, we can assume that the blackbody curve is symmetrical in the visible region around 550 nm (it is actually skewed a little to the red, but this will introduce only a small error). Then an average lumens per watt conversion factor can be calculated from the luminous efficiency curve by summing the points at 10 nm intervals and dividing by the number of intervals (in this case, 30). The value is about 235 lm/W, averaged over the visible region. The flash delivers 0.07 lumen to each square centimeter of object, so the visible irradiance per square centimeter received from the flash with this illuminance is roughly 0.07 lm $\div$ 235 lm/W, which is equal to 3.0×10^{-4} W visible light/cm^2. The amount of power in the ultraviolet light is 15% of that in the visible, or $0.15 \times 3.0 \times 10^{-4}$, which is equal to 4.5×10^{-5} W/cm^2. Over the duration of the flash, 1/100 of a second, the ultraviolet energy arriving at the object per square centimeter is thus 4.5×10^{-7} J/cm^2.

Next, the approximate number of ultraviolet photons can be determined and compared to the approximate number of molecules present. Equation 1.3 can be used for this purpose if the wavelength of the ultraviolet energy is known. If an average near-ultraviolet wavelength of 380 nm is assumed, the number of ultraviolet photons from each flash arriving per square centimeter of object at a distance of 1 m is approximately (4.5×10^{-7} J/cm^2) $\times$ ($380 \div 2.0 \times 10^{-16}$) photons/J, or about 8.6×10^{11} photons/cm^2. That is, each flash from this point-and-shoot camera delivers slightly less than 9×10^{11} ultraviolet photons to every square centimeter of object at a distance of 1 m. A moderate quantum yield for a photochemical reaction might be 0.00001, or 10^{-5}, so about nine million molecules per square centimeter will undergo a chemical change if all the photons are absorbed. This may seem to be a very large number, until you consider the mass of this much dye. For example, nine million molecules with a typical molecular weight of 450 would weigh only 6.5 billionths of a microgram. If you compare this to the size of a paint chip weighing only a few micrograms, you realize that the largest possible change per flash caused by near-ultraviolet light from this particular flashlamp—for this quantum yield—is negligible.

The second example is at the opposite extreme: the xenon-arc lamp in the Atlas Weather-Ometer when used with filters to simulate Miami peak daylight during an artificial aging experiment. The information provided with the instrument specifies that when the lamp is operated at 2500 W, 31.3 W of power between 320 and 400 nm are incident per square meter of sample surface or, in terms of square centimeters, 0.0313 W/cm^2 (Atlas Electric Devices Co. 1996). There is no attenuation of this light by additional filtration. Analysis of the spectral curve given in the manual by 10 nm intervals through the ultraviolet region, and the use of equation 1.4, indicates that 5.6×10^{16} photons of ultraviolet light per second impinge on each square centimeter of sample. If, again, the quantum yield for chemical change is 10^{-5}, then 560 billion molecules will react per second. For a dye of molecular weight 450, one microgram will have reacted in a 40-minute experimental exposure. This is an extreme case of an exposure that is not acceptable to conservators and is considered here for a hypothetical material that is only moderately likely to undergo photochemical change. One assumes that art objects of this moderate stability are not left outside in Miami noonday sun.

The third example is a xenon strobe lamp manufactured for studio photography. It is rated at 1000 volts (V), and it has a Q3 synthetic quartz housing. The output of this assembly is adjusted to that of a blackbody with a color temperature of 5300 K by deposition of a dichroic film on the housing. The spectral information supplied by the lamp manufacturer to the system fabricator, however, is only for lamps rated at 450 V and 2000 V, probably for proprietary reasons. The lamp outputs differ mainly in the infrared, which will not significantly alter the photochemical reactions caused by the light energy. Also, the numerical data supplied by the system fabricator are for total power output. Thus, the amount of ultraviolet light energy emitted by the 1000 V lamp must be obtained by interpolation of the data given for the other two lamps. The results indicate that the total energy output per flash is 46.5 J; of this, about 22 J are in the visible region, between 400 and 700 nm, and 0.5–0.7 J per 10 nm interval is in the ultraviolet between 360 and 400 nm. The rest is emitted at wavelengths between 700 nm and approximately 1000 nm and is not at all likely to cause a photochemical reaction. There is negligible energy output below 360 nm because of the Q3 housing; however, the maxima of spectral power curves available for evaluation are at 480 nm, corresponding to a blackbody curve for 6000 K. This indicates that the data are for the lamp before the dichroic filter is added, an assumption confirmed by the fabricator's technical services department. Given this information, the calculation needs to be modified, substituting a blackbody curve of 5300 K for the 6000 K and recomputing the number of ultraviolet photons. The result is a total output of approximately 25×10^{17} photons per flash between 360 and 400 nm.

For the purpose of this calculation, the lamps are estimated to be 1 m from the object and the reflectors in the lamp assembly housing are assumed to direct the flash so an area of approximately 10 m^2 is illuminated. (This is a generous estimate, equivalent to a square slightly larger than 3 m on a side, or a circle roughly 3 m in diameter, and it may lead to a slight underestimation of the photochemical effect.) In contrast, doubling the distance from the flashlamp to the object reduces the incident energy per unit area by a factor of 4. For this example, approximately 25×10^{12} ultraviolet photons would strike each square centimeter of the object per flash. A quantum yield of 10^{-5} implies that 2.5×10^{8} molecules could react per flash per square centimeter. Typically, four lamps are used in a studio, so one billion molecules may undergo photochemical change per flash. Again, this large number should be considered in relative terms; for example, one square centimeter of a piece of paper may contain roughly 3×10^{19} monomer units of cellulose, sizing, or filler material. Therefore, fewer than one molecule (or monomer unit) per ten billion is affected by each flash from the four studio lamps. More than 100 million flashes would be required to cause photochemical change in about 1% of the paper in this example, using a reasonable quantum yield of 10^{-5}. The very large number of flashes needed to reach this probably unacceptable level of degradation suggests that for these particular conditions, conservators need not worry about exposure of this type of material to the near-ultraviolet light emitted by xenon strobe lamps in the photography studio.

The light source in a photocopier or scanner may also be a xenon lamp, which is presented here as the fourth example. Ultraviolet filters are not usually employed in these devices; the ultraviolet blocking is provided by the lenses and glass platen between the source and the original to be copied; thus, near-ultraviolet wavelengths greater than 320 nm are likely to reach the object. Evaluation of the spectral output of a typical xenon source used in the late 1980s indicates that, per copy, the total visible radiation is roughly 5×10^{16} photons/cm^2; about one-third of this visible-light energy is in the 400–500 nm region. Approximately 3×10^{15} photons/cm^2 of wavelengths below 400 nm also reached the original. This represents slightly less than 10% of the total energy of the radiation, because the energy of the ultraviolet photons is somewhat higher than that of the visible-light photons. Making a single copy of an archival document with this particular source, for example, could expose the object to more than 100 times as many ultraviolet photons, and almost 100 times as many visible-light photons, as a studio strobe flash exposure. Also, the shorter the near-ultraviolet wavelengths, the higher the quantum yields may be in some materials, making photochemical change more likely in these substances.

This numerical example is based on data for a single lamp at a particular stage in its useful lifetime. It is important to keep in mind that the output from these photocopier light sources can vary, depending on a number of factors, including the age, type, and supplier of an individual lamp, as well as its sensitivity to fluctuations in voltage. The numbers calculated here should be regarded as rough estimates for the purpose of providing illustrative examples.

The ultraviolet outputs of the four different examples of xenon light sources discussed here are summarized in table 1.2. The major distinguishing features are listed, as well as the assumptions made regarding the individual lamps that permitted completion of the calculations. The total outputs of visible-light photons, from 400 nm to 700 nm, have been approximated in a similar manner and are also included.

Tungsten and tungsten–halogen filament lamps

Pure tungsten sheet metal emits somewhat more light in the visible region than would be expected for its equivalent blackbody temperature (Phillips 1983a); however, most of the total energy emitted by tungsten filament lamps (fig. 1.5) is in the infrared. Only about 2% of the energy emitted from a tungsten filament at wavelengths below 700 nm is in the ultraviolet. When the filtering effect of the glass housing is included, this drops to roughly 1% in the near ultraviolet (320–400 nm). Thus, the amount of near-ultraviolet energy emitted by tungsten light sources is often within the museum guideline of being no greater than 75 μW/lumen.

The use of halogen gas in tungsten lightbulbs to increase the lifetime of the filament approximately doubles the relative amount of ultraviolet light emitted per lumen. A few decades ago, another combination lamp became popular for light exposure studies: the mercury–tungsten lamp, marketed as the Microscal lamp, which was designed to simulate daylight filtered through window glass. Feller (1994, 93–95) presents a spectral power output curve for this source. Between 300 and 600 nm, this closely

Table 1.2
Calculated approximate exposures per flash for xenon sources

Xenon lamp	Handheld camera, color corrected	Studio strobe, 4 lamps	Typical photocopier, late 1980s	Atlas Xenon Weather-Ometer Ci 4000
UV wavelength range (nm)	360–400	360–400	340–400	320–400
UV photons/cm^2	9×10^{11} per flash	1×10^{14} per flash	3×10^{15} per copy	5.6×10^{16} per second
Visible photons/cm^2	9×10^{11} per flash	1×10^{15} per flash	4×10^{16} per copy	5.5×10^{17} per second
Assumptions	1. Object distance = 1 m 2. Average lm/W in visible region	1. Area 10 m^2 is illuminated 2. Dichroic filter shifts blackbody temperature to 5300 K	1. Exposure per copy, not per flash 2. UV blocking only by optical components and platen	1. Exposure per second, not per flash 2. Filters in place to simulate Miami peak daylight

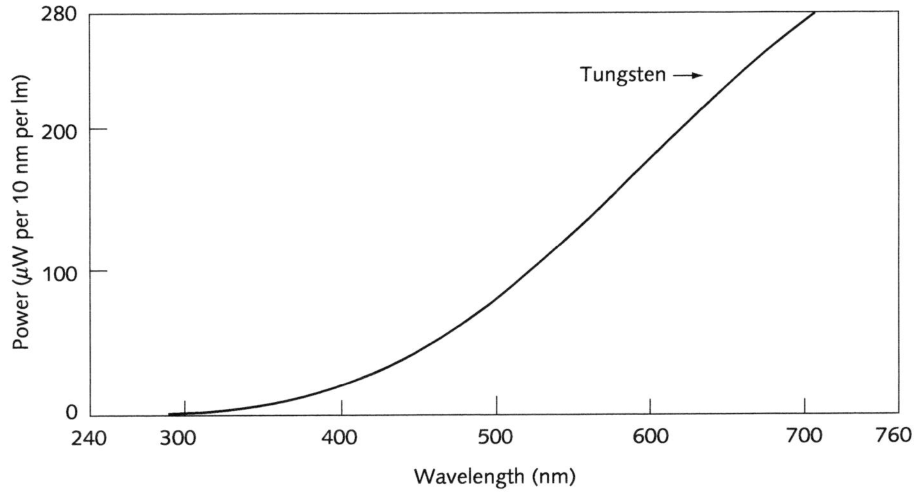

Figure 1.5
Relative spectral output from a tungsten incandescent lamp. (From Thomson 1986, fig. 59a, by permission of Butterworth/Heinemann.)

resembles the output from a mercury-vapor lamp, with very strong mercury lines superimposed on a low background. However, the relative output between 600 and 750 nm, which is 40% of the total, is much greater than that from a mercury-arc lamp. There is a local maximum at approximately 660 nm. Very roughly, about 15% of the energy emitted is in the ultraviolet region between 300 and 400 nm—very little of which is below 360 nm; and a similar amount is in the blue between 400 and 500 nm.

Tungsten–halogen light sources may also be used in photocopiers. Evaluation of the spectral output from one of these lamps used in the 1980s indicated that about 1.5% of the energy per scan is in the near ultraviolet, corresponding to just over 10^{15} UV photons/cm^2 per copy for this source. The output increases with wavelength; almost two-thirds of the energy is at wavelengths greater than or equal to 600 nm. Again, these values can vary from lamp to lamp and with the age of the lamp.

Neevel (1994) also investigated the ultraviolet output of several photocopiers in use in the Netherlands. All but two of the instruments evaluated had tungsten–halogen light sources; most were medium-volume photocopiers. The near-ultraviolet exposure per copy was inversely related to the time taken to make the copy; that is, the high-speed, high-volume photocopiers delivered the lowest ultraviolet exposures per copy. Neevel rated the photocopiers according to the approximate number of copies that would expose the object to as much ultraviolet light as an art object would receive if illuminated by a tungsten light source at approximately 50 lux for one hour. In one case, the exposure varied by a factor of 10 from the center to the edge of the object. Some of the photocopiers delivered the equivalent of an hour's worth of near-ultraviolet energy from a 50 lux tungsten source in ten or fewer copies. Multiple copying on a machine with one of these "worst case" sources could quickly shorten the display lifetime of a sensitive object.

Fluorescent lamps

Fluorescent-light tubes are designed to emit a variety of spectra for different purposes; there are "daylight" tubes, plant-growth lights, germicidal lamps, "cool-white" or "warm-white" lights, and so on.

This variation is accomplished by using different phosphorescent substances to coat the inner wall of the tubes, modifying the gas pressure and composition, and even altering the nature of the glass housing itself. These sources typically emit a continuum of light from the near ultraviolet through the visible region, with a maximum at about 600 nm. All fluorescent tubes contain mercury, the emission lines of which are superimposed on this continuum (fig. 1.6). The relative amount of light in the ultraviolet will be strongly influenced by the lamp specifications and also by the nature of any ultraviolet filtering that may be present.

In most photocopiers, the glass housing of the lamp and the glass platen will not transmit light of wavelengths less than about 320 nm. There may not be any other light filters present. A particular fluorescent lamp used as a photocopier light source in the late 1980s irradiated the original being copied with a total energy of approximately 21 millijoules (mJ). This lamp had very prominent bands in the visible region at 435 nm and 610 nm, which accounted for 13 and 17% of the energy, respectively; the lamp delivered approximately 6×10^{15} and 8×10^{15} photons/cm^2 per copy at those wavelengths. If two copies were made of an object with a fugitive dye at moderate coverage, and the dye was sensitive to these wavelengths (for example, 10^{16} molecules/cm^2 and a relatively high quantum yield of 0.001, or 10^{-3}, at either wavelength), up to 0.1% of the molecules might be degraded. This is roughly equivalent to using up one-tenth of the total allowed photoflash or reprographic light exposure for the lifetime of this object. Objects with low coverages of highly fugitive colorants probably should not be copied on a photocopier or scanning device with this type of light source. Because the properties of individual sources may vary significantly, the calculation presented here is a cautionary example only.

Comparison of the spectral power distributions of the examples of xenon and fluorescent light sources in photocopiers is instructive. The output of the fluorescent source is roughly the same below 400 nm as that of the xenon lamp. However, the fluorescent source emits as many photons in the 400 and 430 nm regions as the entire visible out-

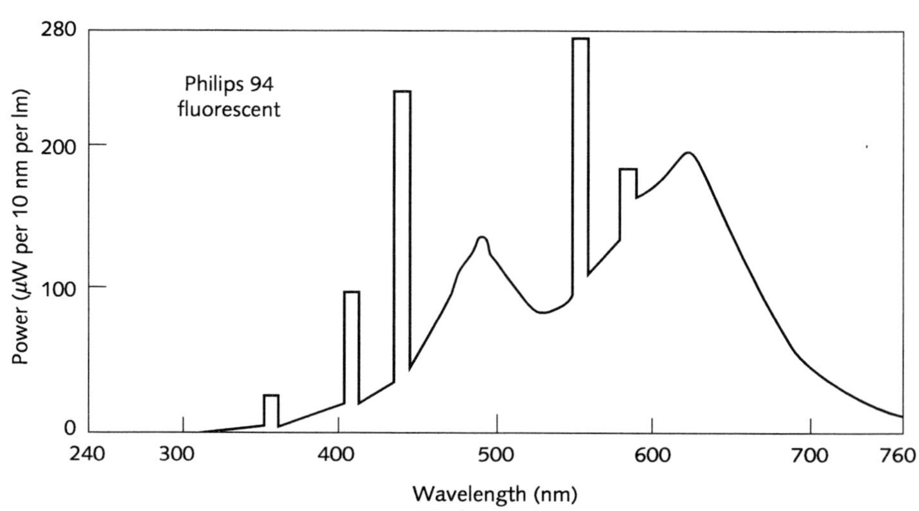

Figure 1.6
Relative spectral output from one type of fluorescent lamp, the Philips 94. (From Thomson 1986, fig. 59c, by permission of Butterworth/Heinemann.)

put of the xenon source. Thus, materials that absorb blue light will have a higher risk of being damaged by the fluorescent than by the xenon source.

Two different types of ultraviolet fluorescent lamps have also been used as light sources in photochemical and lightfastness research (Jagger 1977). The BL black light emits a broad band of radiation from 300 to above 450 nm, with a maximum at 365 nm; it puts out slightly more energy above 400 nm than it does between 300 and 350 nm. Approximately one-half of its emission is between 350 and 400 nm. The other ultraviolet fluorescent light is the BLB lamp. It has a narrower emission band with a maximum at 350 nm. It emits as much energy between 350 and 380 nm as does the BL light despite having only about two-thirds the total output. It is difficult to relate research results obtained with this lamp directly to the effects of photoflash and photocopier lights because of this very different spectral distribution.

Mercury-arc and carbon-arc lamps

Ultraviolet-light photography is sometimes necessary to evaluate the condition of oil paintings and other objects for conservation purposes. A mercury-arc lamp is usually used as the (near-)ultraviolet light source for this photography. It is thus appropriate to examine in detail the effect of light energy from this type of source on museum objects. The mercury lamps employed are normally high-pressure arcs, which emit a continuum background as well as a sharp, energetic line spectrum characteristic of the element mercury (fig. 1.7). A significant amount of photochemical research has been done with mercury-arc light sources, so considering the output from mercury lamps will also simplify evaluation of the results of the investigations reviewed in this publication.

A typical 100 W mercury-arc lamp may emit roughly 650 μW of 300–400 nm light per lumen, with approximately 75% of this power in a major line at 365 nm. The output in the visible region is roughly 2500 lm. This means that about 1.6 W of near-ultraviolet light are emitted, and approximately 1.2 W of this are at 365 nm. Filters usually used

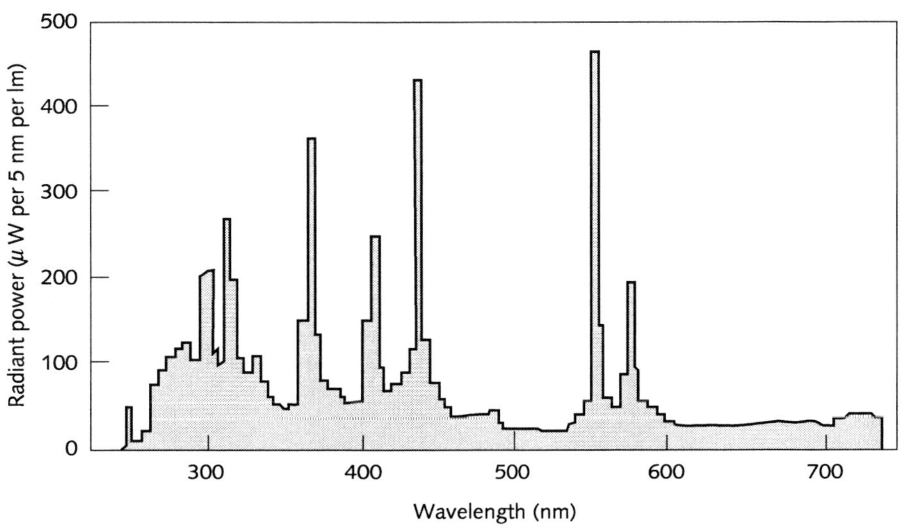

Figure 1.7
Relative spectral output from a high-pressure mercury-arc lamp. (From Suppan 1994, fig. 7.3c, by permission of the Royal Society of Chemistry.)

with these lamps in the conservation photography laboratory will eliminate visible light, and wavelengths less than about 320 nm may also be removed by filtration; however, as much as 1.4 W of power in the near ultraviolet are likely to remain in the emitted light beam. For a one-minute exposure, this corresponds to approximately 1.5×10^{20} near-UV photons. If these are all incident on a 0.3 m^2 (3000 cm^2) area, then 5×10^{16} photons of near-UV light strike each square centimeter of the object during each exposure. If all of this light energy is absorbed by materials in the object, and if a quantum yield of 10^{-5} is appropriate, 500 billion molecules would undergo a photochemical change. Although this number is much higher than those calculated for the xenon flash-lamps, it is still a small percentage of the total number of molecules likely to be present in 1 cm^2 of the object. Conservators should be concerned about the effect of this ultraviolet-light exposure if the probable quantum yield for a photochemical change is significantly greater than the 10^{-5} assumed here. Also, if filters used with this light source do not effectively block wavelengths of light below 320 nm, there is a greater likelihood of photochemical damage due to the higher absorbances and the higher quantum yields for photochemically induced reactions at the shorter ultraviolet wavelengths.

A significant portion of the visible-light energy emitted by mercury-arc lamps is located in a few strong, narrow bands centered at 405, 436, 546, and 578 nm. An unfiltered, high-pressure 100 W lamp emits approximately 8, 20, 26, and 28% of the visible energy output at these wavelengths, respectively. The corresponding quantum outputs from this particular lamp are roughly 1×10^{18}, 3×10^{18}, 4.5×10^{18}, and 5×10^{18} photons per second. Depending on the lamp housing, the reflectors used, and the extent of visible-light filtering, the actual number of photons striking the object per square centimeter could exceed the coverage of colorants. If the colorant absorption bands coincide with any of these wavelengths, the possibility of photochemical reaction could be significant.

Those referring to the older photochemical literature for the first time should note that categories of ultraviolet light identified as UV-A (about 325–400 nm), UV-B (280 nm to about 325 nm), and UV-C ($\leq$280 nm) were not in general use prior to the 1970s. In fact, the division between near-ultraviolet and visible light, now taken to be 400 nm, was not always distinctly made. Some articles refer to all light of wavelengths above 320 nm as "visible," and others may mention "UV spectroscopy" in general while including visible wavelengths shorter than about 600 nm (e.g., studies done using a mercury-arc lamp with a glass filter to block the 257 nm and 313 nm bands, or in the absence of ultraviolet-blocking filters, respectively). The actual spectral output of light-source setups used in the experiments must be confirmed in all cases before the information given in summary statements can be used to evaluate the light stability of any particular material.

Carbon-arc light sources are probably not used in museums today. However, some of the older lightfastness research was performed with these lamps (early versions of the Atlas Weather-Ometer and Atlas

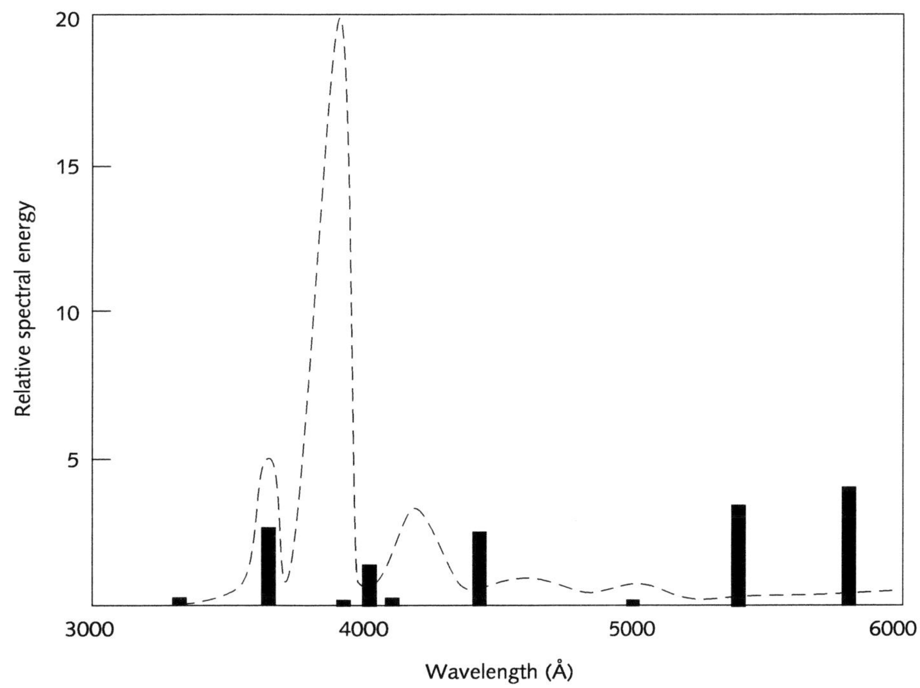

Figure 1.8
Relative spectral output from a carbon-arc lamp. (From Giles and McKay 1963, 529, by permission of the *Textile Research Journal.*)

Fade-Ometer have carbon-arc lamps), and they were occasionally employed in early photochemical studies. Some relevant attributes of these sources are summarized here.

The emission spectrum for the type of carbon-arc lamp commonly used in accelerated fading tests is shown in figure 1.8. This spectrum vaguely resembles that of sunlight in the wavelength range 350–430 nm, but it lacks significant energy in the rest of the visible region. The light source was useful for research on materials that absorb in the region where this type of lamp has considerable output—that is, in the near ultraviolet and blue range—but it would not have provided informative results for substances that absorb green or red light (or both). About 60% of the total energy emitted by this carbon-arc lamp was in the major band centered at approximately 395 nm; 10–12% was in the ultraviolet band centered at 365 nm. Roughly 10^{16} photons/second were emitted in each of these two major bands. From 10 to 12% of the output was also in the blue band between 400 and 430 nm. Results obtained with this type of lamp should be evaluated for their relevance accordingly.

Another point to consider about carbon-arc light sources is that they were widely used as photographic studio lighting, as graphic arts light sources, as spotlights, and as movie projector lamps in the first half of the twentieth century (Phillips 1983a). Collections that were frequently displayed and photographed during that period, as well as older films that were repeatedly projected, have been extensively exposed to the blue and near-ultraviolet light from carbon-arc lamps (and the gaseous and particulate pollutants they emit). The materials in such objects may be expected to have undergone some photochemically induced changes if they absorbed strongly in the 350–430 nm region.

Filters Used with Light Sources

The transmission characteristics of filters used with light sources can be distinguished by categorizing them as band-pass or cutoff filters. Physically, these may consist of films of material adhered to glass or plastic, glass material incorporating absorbing dycs, or thin metallic layers vacuum deposited on quartz or glass to exacting thicknesses (dichroic filters).

Catalogues of filter transmission spectra are available from producers of color glass filters and of cutoff filters for optical equipment used in research (fig. 1.9). In contrast, obtaining complete specifications for the filters and windows incorporated into photoflash lamp assemblies has proved very difficult. Many filters for photographic light sources are proprietary, and manufacturers have not been willing to supply precise information about the spectral transmission or blocking characteristics of these products. Some camera makers have also been unable or unwilling to provide information on the use of filters on the xenon flashlamp assemblies in their large-volume-production cameras for amateurs. Manufacture of these components is contracted out, and suppliers and processes may change frequently. In still other cases, in which dichroic filters are deposited directly onto individual lamp housings, the manufacturer will supply information only on a particular lamp for which the customer requests standardization and specifications, which are costly to obtain. Specifications tend to vary, and manufacturers want to avoid liability for performance claims on individual product items that have not been inspected or measured. This situation has made impossible the accurate calculation of spectral output characteristics for some light sources, particularly with respect to ultraviolet light emission. In the absence of information to the contrary,

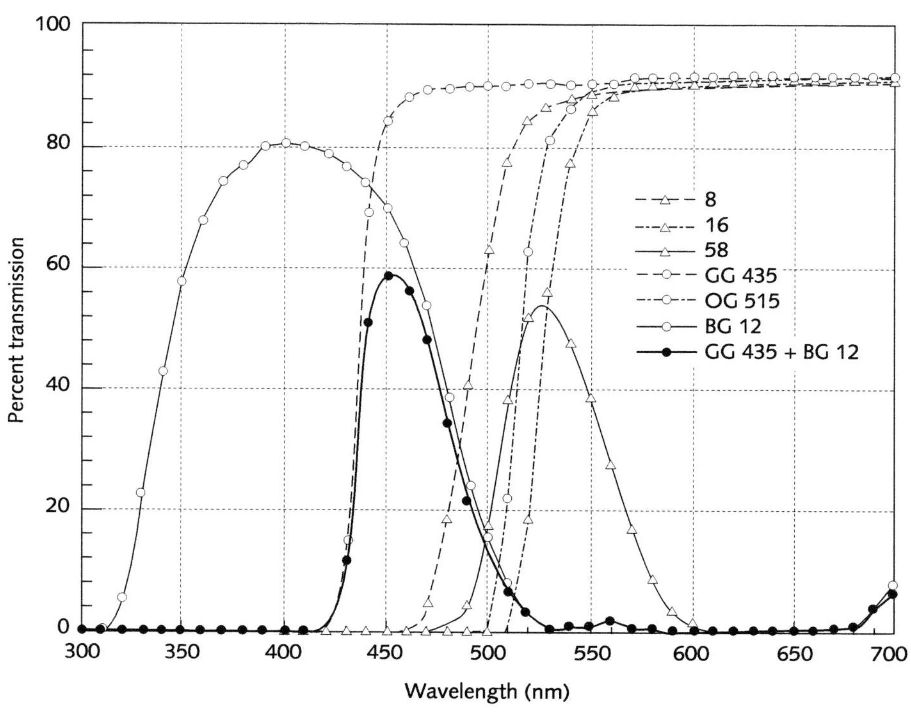

Figure 1.9
Transmission spectra of several glass cutoff filters (dashed lines) or band-pass filters (solid lines), manufactured by Schott (circles) or Kodak (triangles). A narrower band of light can be selected by the combination of a sharp cutoff filter and a broad band-pass filter, as indicated by the thicker solid curve. (Data from Internet site of BES Optics Inc.—a distributor for Schott filters—and from Kodak 1990.)

it may be wise to assume that there is no blocking of wavelengths greater than about 320 nm in the light emitted by photoflash or reprographic light sources.

To obtain a rough estimate of the transmission of the glazing on built-in flashlamp housings of point-and-shoot cameras, two disposable cameras (Fuji and a generic camera recycled for Recreational Equipment Inc. [REI]) were dismantled. The rectangular plastic glazings on these flashlamps are ribbed to diffuse the xenon flash, and both looked slightly yellow when viewed at an angle on a white tissue. For testing, each glazing was masked with matte-black electrical tape to avoid scattering by the ribbed plastic, except for a strip approximately 1 mm wide; the cuvette holder of a Beckman DU-62 spectrophotometer was similarly masked. Then the transmission of the unmasked area of each glazing was measured from 300 to 550 nm at 5 nm intervals with the instrument calibrated against air at each wavelength. The data are graphed in figure 1.10 and indicate that ultraviolet light is effectively blocked by both of these flashlamp window coverings.

Ultraviolet-blocking plastics in vitrines and in glazings on museum objects could also block much of the near-ultraviolet light from flashlamps used in galleries. Figure 1.4 shows that UF-3 plastic glazing would prevent most of the near-ultraviolet wavelengths emitted by built-in camera flash sources from reaching objects glazed with this material. This conclusion would also apply to objects displayed in vitrines made of these materials. OP-3 acrylic transmits less of the near ultraviolet than does UF-3 and would be expected to block almost all the ultraviolet emitted by xenon flashlamps.

Figure 1.11 shows the transmission characteristics of several other materials that may be used as lamp housings or placed between art

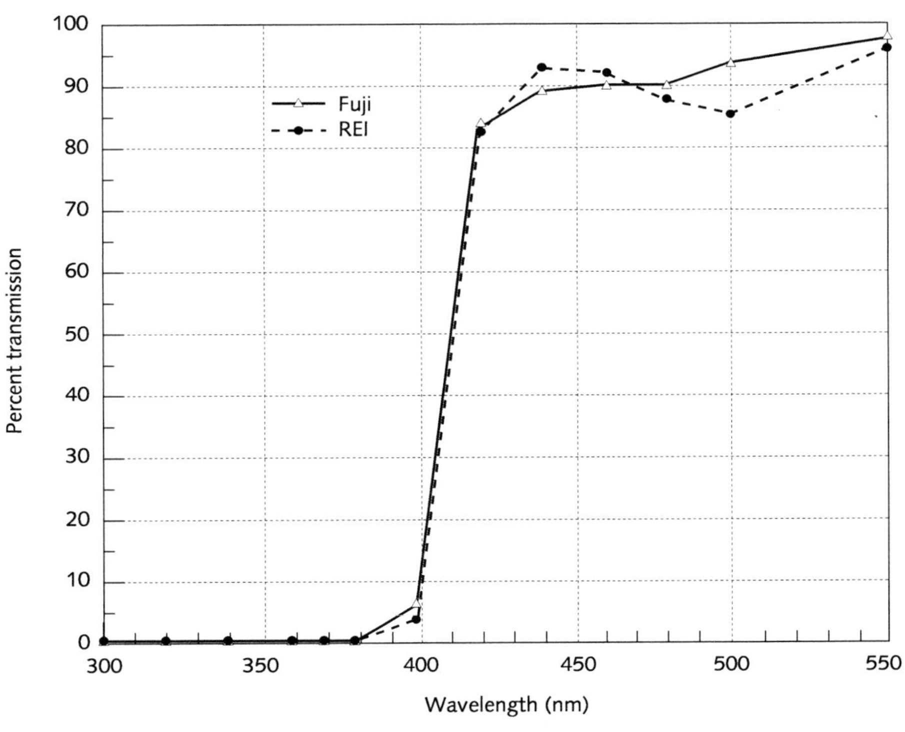

Figure 1.10

Transmission curves of plastic windows of xenon flashlamp housings in two disposable cameras. The curves were measured at 5 nm intervals with a Beckman DU-62 spectrophotometer.

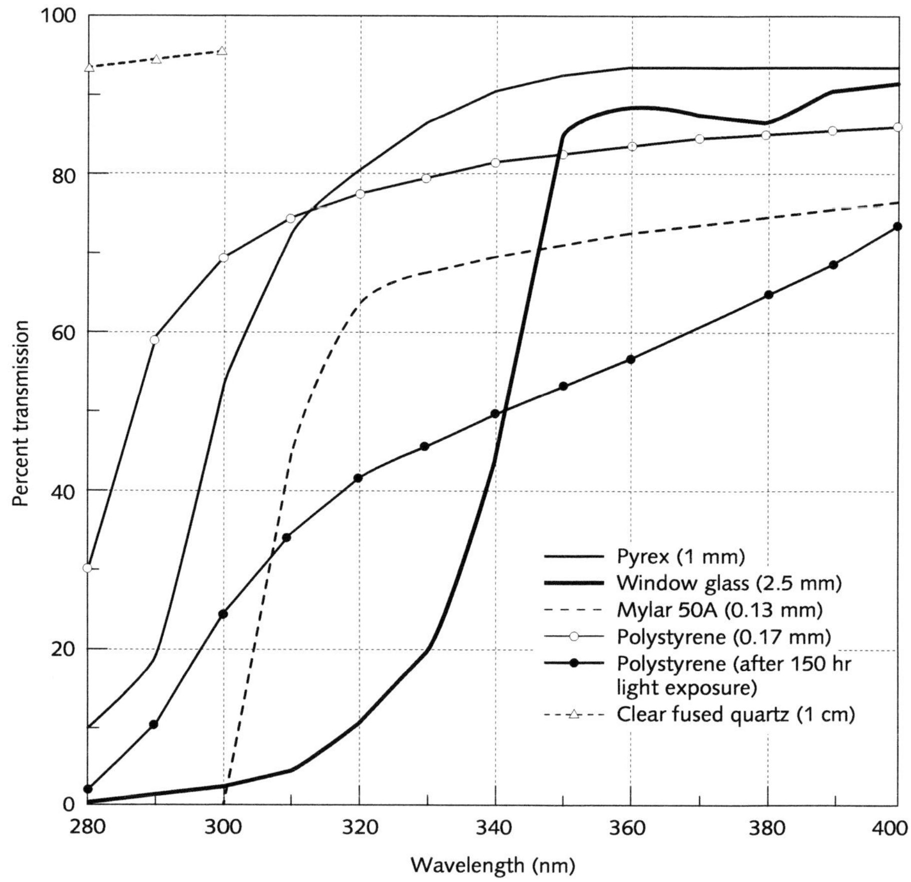

Figure 1.11
Transmission spectra in the near-ultraviolet region of various materials used for lamp housing, glazing, or protection of art and archival objects. The transmission of some materials in this wavelength region may change after exposure to ultraviolet light in this energy range. (Data from Jagger 1967, 1977.)

or archival objects and light sources. When new, polystyrene film does not have significant UV absorption above 300 nm; after aging in light, however, it absorbs 30–60% of the near ultraviolet above 320 nm. Of the materials shown, window glass is the most effective UV-blocking agent for higher energy, UV-B ultraviolet light.

Crews (1989) determined the transmission characteristics of four ultraviolet-filtering materials—a UF-3 acrylic sheet and three flexible polyester films—as well as their effectiveness in decreasing the fading of several natural dyes on wool fabric. The samples were exposed to a xenon-arc lamp in a fading device with and without the filters. As might be expected, an amber-tinted film that absorbed significant amounts of blue light was the most effective of the four materials in preventing sample fading. It was also the least stable of the filters tested. As total light exposure accumulated, this amber filter transmitted increasing amounts of light between about 370 and 500 nm. This is expected because it absorbed the most near-ultraviolet and visible blue light and would have undergone some photochemically induced change.

Considerations for the Calculations

Important assumptions

To avoid misinterpretation, the contexts in which data are evaluated and the results applied to various situations must be kept in mind.

Assumptions and estimates, as well as the conditions under which the data were originally acquired, can supply this context. The importance of recognizing the consequences of assumptions made in the evaluations cannot be overemphasized. This is illustrated by comparing the conclusions Michalski posted on the Internet (Conservation Distribution List 1996) with those of Neevel (1994) regarding flashlamps used by professional photographers.

After measuring the ultraviolet output of various European lamps, Neevel calculated that in about 10 flashes some of them would deliver an amount of ultraviolet light equivalent to that radiated in one hour by 50 lux incandescent gallery lighting. Michalski, assuming that the ultraviolet output of the flashlamps could be neglected, approximated that 225 flashes from two professional flashlamps would be required to deliver the equivalent of one additional hour of light exposure to objects displayed at 50 lux under incandescent gallery lights. This is more than forty times Neevel's estimate. The discrepancy in this example is due to Michalski's decision to assume that the ultraviolet output of the lamps could be ignored, and not to the use of different acceptable exposure increases, different gallery lighting levels, or arithmetic errors. Michalski made this assumption, it is interesting to note, despite the results of his earlier work leading to the development of the light-exposure slide rule (Michalski 1987). In presenting this work, Michalski summarized evidence that the relative sensitivities of various materials to ultraviolet light varies compared to their sensitivities to visible light and that ultraviolet sensitivity cannot always be predicted with reference to the latter.

Two assumptions routinely made in performing the evaluations presented in chapters 2 and 3 were that near-ultraviolet wavelengths greater than or equal to 360 nm have an effect, unless calculations suggest otherwise, and that the reciprocity principle is obeyed, unless the literature data show otherwise for a particular material (materials that do not obey this principle are identified in the relevant subsections of chap. 2).

The assumption by Neevel and Michalski that flash and reprographic light exposures be allowed to consume 10% of the display lifetime is not used here. Instead, the total allowable flash exposure is assumed to be only 1% of the display lifetime. *Total display lifetime* is defined as the exposure time required to cause a perceptible change in appearance or an unacceptable change in a physical property related to the integrity of the object. Thus, the low value of 1% essentially prevents flash exposures from having any impact on an object's display time. In a particular museum situation, those responsible for the care of collections would need to decide on a satisfactory percentage in consultation with curators and administrators when risk assessment policy is established. The relative importance of consuming some display lifetime for an immediate purpose, thereby decreasing future display time (or increasing future risk of fading during display), can be compared to the immediate benefits of permitting photography or reprography. Collections managers will have to decide which course of action is best for their particular situations. The lower level of risk—permitting only 1% of total display lifetime to be consumed by flash exposure—may be very restrictive in some instances. For example, for the worst-case flashgun in Neevel's survey, only one flash per day of exhibit would be allowed for objects displayed at 50 lux; fewer exposures would be permitted for more-sensitive objects. These restrictions would preclude gallery photography and would limit studio or conservation photography over the entire course of the multiple-exhibit lifetime of a very sensitive object.

Other assumptions were made in particular cases in which specific numbers not included in the literature were required to complete calculations of allowable numbers of flashes. For example, several approximations and estimates were used and order-of-magnitude (power-of-ten) numbers were considered for the calculations of xenon lamp outputs described earlier in this chapter. This approach was originally taken in an early study that considered the lightfastness of mixtures of dyes on fibers (Hemmendinger 1972), and it remains useful today. The procedure is appropriate because of the large numbers of photons involved and the even larger numbers of molecular entities in the objects. This estimation technique can provide a qualitative idea of possible effects. Order-of-magnitude estimates were also used here in evaluating literature data on the photochemistry of materials found in art and archival objects.

In the absence of specific photochemical data, calculations of flash effects were performed with the help of other information. In particular, lightfastness ratings and fading kinetics reported with reference to Blue Wool standards were assumed to be suitable bases for the order-of-magnitude calculations. These data are of considerable importance because most of the detailed results available for dyed and pigmented polymeric materials have been published in this context. Because of the importance of the assumptions inherent in the use of Blue Wool standards data—and some additional problems with their use—these assumptions are discussed separately later in this chapter.

Approximations and estimates

To express the spectral outputs of light sources discussed here, approximations were made of the numbers of photons at various wavelengths. Similarly, the numbers of possible reactive molecules in the works of art are expressed as rough estimates. For example, in one of the examples cited earlier, 1 cm^2 of a paper weighing slightly less than 10 mg is reasonably estimated to contain more than 10^{19} monomer units of cellulose or sizing. A similar estimate for 1 cm^2 of a gelatin layer about 10 μm thick on a black-and-white photographic print suggests that there are approximately 6×10^{18} amino acid units/cm^2, if an average formula weight for the amino acids in gelatin is taken to be 105.

Estimating the number of pigment molecules per square centimeter for typical objects is more problematic because of the wide range of thicknesses possible in pigment application and in the percentage of binder or other components used. As an example, Daniels (1995) assumes that 30% of the material by volume is binder and that the pigment is applied in a layer 0.2 mm thick. The volume of pigment itself in a 1 cm^2 area in this case is $1 \times 1 \times 0.02 \times 0.7 = 0.014$ cm^3. Further, if the density of the colorant is assumed to be 1.5 g/cm^3 and the formula weight 300, many more than 10^{19} molecules of colorant will be found in each square centimeter of the paint layer. This number could easily vary by more than a factor of 10 in either direction, depending on the thickness of the layer and the actual composition of the pigment. For a natural dye used in delicate application on a Japanese print or brush painting, for example, a number a few orders of magnitude smaller may be appropriate.

A lightweight, plain-weave, silk test fabric, called *habutae*, has a typical density of 3.1 mg/cm^2. When an average molecular weight per amino acid monomer unit in the silk protein chain is assumed for the purposes of approximation, this lightweight textile is calculated to have a density of 2×10^{19} monomer units/cm^2. Other silks may weigh from one-third more to about twice that before they are sized and dyed, so up to twice as many monomer units may be present in a given area. Heavily weighted silk, which became popular in the last century when manufacturers sold silk by weight and included as much metal as possible to raise the price (van Oosten 1994), may not contain any more protein per square centimeter. In fact, there may be less protein remaining because of degradation.

These calculations of the numbers of molecules or monomer entities per unit area indicate that in most circumstances a considerable excess of target molecules compared to photons from a xenon photoflash source is typical. Another estimate used consistently in chapters 2 and 3 is that essentially all photons of wavelengths that can cause a photochemical change in the object are absorbed. This may be an overestimate, but it is likely to be a slight one, causing underestimation of the allowable number of flashes by less than an order of magnitude.

Blue Wool standards

Some institutions have devised display policies based on the Blue Wool scale of lightfastness. This scale consists of eight different blue dyes on wool fabric, each dye being roughly one-half to one-third as sensitive as the previous one to fading by light. The lightfastness of materials is stated on the basis of the visible-light exposure, usually quoted in lux hours, that causes the material to fade to the same extent as a particular Blue Wool standard. The formulations for materials used for Blue Wool standards in the United States differ slightly from those used in the United Kingdom, but they all meet the ISO performance specifications. The total light exposure that is generally recognized to cause a just-perceptible fade in Blue Wool no. 1, the most sensitive standard on the scale, is 4×10^5 lux hours of visible light (Colby 1992). Therefore, this would be the total display lifetime of an object with a lightfastness equivalent to Blue Wool no. 1. The calculations given earlier (see "Light Sources," pages 13–25) indicate that this limit is reached after twenty display periods of forty 10-hour days at a light level of only 50 lux. If the materials in the object obey the reciprocity principle, more frequent or longer exhibits could be mounted at lower light levels.

The assumption that only 1% of the total display exposure should be consumed by flash (see "Important assumptions," pages 28–30) would suggest that the total exposure from flash should not exceed 4000 lux hours of visible light for an object with sensitivity equivalent to Blue Wool no. 1. The camera flash radiating the least light, with ultraviolet blocked, emits 700 lux of visible light per flash of 0.01 second duration. So it would take approximately two million of these flashes—(4000 lux hours × 3600 seconds/hour) ÷ 7 lux seconds per flash—to consume 1% of the display lifetime of an object with this lightfastness rating. This is equivalent to about one hundred thousand flashes for each of the twenty exhibits described. It represents an upper limit because it is calculated for the least powerful of today's camera flashlamps and for complete ultraviolet blocking. A more cautious limit might be about one-half of this amount, a total that would permit up to 1250 flashes from handheld cameras per day of exhibition. The typical museum exhibit is not likely to receive this much photographic attention from museum patrons. However, highly popular special exhibits or individual objects regarded as icons would probably receive more than this limit if photography were unrestricted or if the objects were displayed more frequently or for longer periods (and these sorts of objects are the ones most often requested for repeated display or loan). On the basis of these considerations, gallery flash photography of very popular objects containing materials as fugitive as (or more fugitive than) Blue Wool no. 1 should not be allowed. For materials that are known to be significantly more sensitive than Blue Wool no. 1, and for objects that are sensitive to near-ultraviolet light, photography might best be limited to studio work essential for documentation and illustration. At the very least, exposure to photoflash and reprographic processes should be recorded for monitoring purposes.

Although Blue Wool standards are rated in terms of sensitivity to visible light, they were originally designed for use in daylight, which includes ultraviolet light. Collections care staff regularly use the standards to monitor radiation of various spectral compositions. Apparently, the response of the standards is often assumed to be the same whether or not near-ultraviolet wavelengths are included. Some authors report results suggesting that this assumption is not appropriate. Schmelzer (1984) and Horie (1990) showed that this premise held under their experimental conditions for Blue Wool nos. 1 and 2, but it appeared not to be applicable to Blue Wool nos. 3, 5, and 6. The results of Levison, Sutil, and Vanderbrink (1987) imply that Blue Wool nos. 6 and 7 do not respond similarly in the presence and absence of ultraviolet light. Giles, Shah, and Baillie (1969) showed that the spacing of sensitivity of the Blue Wool standards is different for light sources having different ultraviolet components. An early investigation by McLaren (1956) indicated that Blue Wool nos. 4 and 5 are less sensitive to blue light of wavelengths 400–460 nm than the other standards. Michalski (1987, fig. 2) summarizes data suggesting that the sensitivity to near-ultraviolet radiation, compared to visible light, is greater for Blue Wool nos. 3, 6, and 7 than for the other ISO Blue Wool standards. Bullock and Saunders

(1999), while undertaking a detailed evaluation of the Blue Wool standards for use as light dosimeters, showed that Blue Wool nos. 1, 2, and 3 faded less when ultraviolet light was removed from the light source by filters. They also obtained data that suggest that Blue Wool no. 3 is relatively more sensitive to UV-A radiation than are Blue Wool nos. 1 and 2. The implications of these observations are considered in chapter 4.

Approach to calculations

If the quantum yield for photochemical change is known, as well as the number of photons with which a particular light source would irradiate the object per exposure and the approximate numbers of reactive molecular entities in the object, then a simple calculation can indicate directly how many flashes will cause change in enough molecules for the deterioration to be perceptible. One percent of this amount is assumed to be the maximum allowable number of flashes. The calculation includes all photochemically active near-ultraviolet and visible wavelengths emitted by the source.

Unfortunately, in most of the examples discussed in chapters 2 and 3, at least one element in this list of quantities was not known or not clearly stated. Frequently the lightfastness value was available instead of the quantum yield. In these cases, 0.01 of the total display lifetime for the relevant Blue Wool fastness rating was used to calculate the number of allowable flashes from a particular photoflash source that emits the corresponding number of photons listed in table 1.2. Often the total dose of light that caused a quantitated appearance change, or the duration of exposure to a particular type of lamp, was stated. To use these data for the calculations, the reported change was compared to the just-noticeable change that determines the total display lifetime of the material. For example, if a total color difference (ΔE) of 20 units had been reported for a pale colorant, and a color change of about 2 units would represent a noticeable change in appearance, the total display lifetime would be assumed to be 2/20, or one-tenth of the reported duration of exposure to the light source used in the experiment. Then flash would be allowed to deliver 0.01×0.1 of the total experimental exposure. Finally, the approximate number of flashes that would add up to 0.001 of that exposure would be calculated.

Frequently, the light source used in a fading study considered here included some photons of wavelengths between 320 and 365 nm, or the relative amount of radiation between 360 and 400 nm was greater than that emitted by filtered or color-corrected camera and studio xenon flashlamps. In these cases, the calculation was based on the number of flashes from the photoflash source required to deliver the total number of photons between 360 and 400 nm that would have caused 0.01 of an unacceptable change. The order-of-magnitude approximations obtained under these circumstances, and reported in chapter 2, are probably underestimates because of unknown contributions to changes in the experimental samples by the higher-energy ultraviolet photons. They are identified as very approximate, or the assumed presence of significant amounts of near-ultraviolet radiation of wavelengths greater than

360 nm is noted. Sometimes the composition of the light was not clearly stated in a report. If other experimental details suggested what type of light source was used, standard data on that type of source were employed to attempt an order-of-magnitude calculation.

In general, the calculations of allowable flashes are cautious estimates. They attempt to suggest an exposure limit calculated to ensure negligible deleterious effects over the display lifetime of the exposed objects. As such, they are designed for situations in which very low risk from exposure to photoflash and reprographic light is desired.

Notes

1. See, for example, Clayton (1970, chaps. 2, 3); and Suppan (1994, chap. 2).
2. These and other properties of light are also discussed in basic texts.
3. See, for example, Brill (1980), Rabek (1987a), and Suppan (1994, chaps. 3, 4).
4. Other units have been used in the past for both radiometric and photometric quantities. Luminance values, especially, have been quoted in a wide variety of units, some of them not easily converted to the current standardized systems. Fortunately, luminance data are not likely to be essential for evaluation of light exposures to which objects in collections are subjected.
5. The blackbody concept and color temperature are discussed more fully by Phillips (1983a) and Schneider (1997).
6. K is the temperature in kelvins; to convert to degrees Celsius, subtract 273 from the temperature in kelvins.
7. A lux is defined as lumen per square meter (lm/m^2) (see table 1.1).
8. Lamps with a Q3 housing coated with a dichroic filter that blocks additional near-ultraviolet and blue wavelengths are used in the photography studio at the Los Angeles County Museum of Art.

Chapter 2

Results of the Literature Search

The findings of the literature search are presented in this chapter, categorized according to types of materials. In each section, relevant background information is briefly summarized, followed by detailed results for some of the more light-sensitive materials. Data evaluations and approximations for other substances are also summarized here.

This chapter contains a wealth of information on a wide range of materials. To help conservators and collections managers locate specific information more efficiently, data on similar topics are grouped together, and key words are highlighted with italics. Whenever possible, extrapolations are made from the literature data to indicate the possible effect of photoflash and reprographic exposures on art and archival materials. Since the extrapolations made and suggestions offered here are necessarily constrained by the author's operating assumptions and by the limits of the original studies, the results presented should be taken as guidelines only and not as hard-and-fast conclusions. If a useful observation or recommendation for the conservator, curator, or collections manager can be made on a particular material, it is given in a "Conservators' Note." Chapter 3 is a companion to this one. It contains additional technical details of the data and calculations presented here, as well as supplementary, related information culled from the literature search. These two chapters are organized in corresponding sections by topic.

Colorants

Colorants are discussed in three separate categories in this section: inorganic pigments, natural organic colorants, and synthetic organic colorants. The colorants in inorganic pigments are generally mineral salts or metal oxides. Historically, natural organic colorants were isolated and purified from biological sources. They have been used as dyes for fibers since antiquity and, when precipitated onto opacifiers, as lake pigments. Synthetic organic colorants have been manufactured by the chemical industry since 1857 (Mills and White 1994a). Although some synthetic dyestuffs and colorants are derived from natural ingredients, the end products are not found in nature as such. Some natural dyes are now also synthesized commercially. These compounds are equivalent to the natural colorants isolated from biological materials and need not be considered separately from the latter.

The responses of colorants to light may be strongly influenced by their surroundings. This can be particularly important in the case of dyed fibers. Therefore data from studies of the lightfastness of colorants in dyed or pigmented textiles and other polymeric materials are presented in a separate section.

Inorganic pigments

Most inorganic pigments are acknowledged to be very light stable, with a few well-known exceptions that include vermilion, lead(II) oxide, lead chromate, historic Prussian blues, and some copper greens (Allen and Edge 1992c). These pigments are discussed in the following sections. Hanlan (1970) was the first to directly address the question of whether exposure to photoflash lamps would cause changes in pigments used in paintings. None of the inorganic colorants he investigated showed a significant change after tens of thousands of flashes from a studio flashlamp.

Vermilion

Feller (1968) showed that the darkening of vermilion is due to conversion of the pigment from one crystalline form to another, and that with appropriate care in following early-twentieth-century recipes, pigments could be formulated that are more than ten times as stable as several commercial varieties. More recently, Daniels (1987) demonstrated that the presence of certain inorganic salt impurities is responsible for the pigment becoming darker when exposed to various light sources, all of which emitted some near-ultraviolet light. Other factors that influence the conversion are the medium; screening of light by filters, films of varnish, or other materials; and the pigment volume concentration, which is effectively another form of screening.

In his experiments, Feller demonstrated that well-made vermilion, which is a moderately light-sensitive colorant, would resist change in a carbon-arc Atlas Fade-Ometer for more than 600 hours, whereas sensitive proprietary brands appeared to be about one-tenth as stable, changing significantly in less than 100 hours. These carbon-arc exposures are much greater than those from individual flashes. If it is assumed that xenon flash photography should contribute no more than 0.01 of the exposure that causes a significant darkening, it can be approximated that sensitive vermilion preparations could be subjected to thousands of flashes (i.e., 10^3–10^4) without deleterious consequences to their appearance. (See the corresponding section in chap. 3 for an expanded discussion of the consequences of these results for photoflash exposures.)

> **Conservators' Note:** *Because the effect of flashlamp exposure is cumulative, some limitation of flash photography may be prudent if an object is known to have poor-quality vermilion.*

Lead(II) oxide

Saunders and Kirby (1996) studied the applicability of the reciprocity principle to light-sensitive pigments and natural dyes. They observed that lead(II) oxide (PbO), yellow pigment massicot, underwent a major color change very quickly. They commented that the change was so rapid that they could not verify the application of the reciprocity principle to this pigment. Of the materials tested, only low-quality paper and litmus showed a greater color change than did lead(II) oxide when exposed to the lower light dose for which data are presented. The slightly elevated temperature (31 °C) of the samples during exposure may have influenced the response of the lead oxide. It cannot have been the major factor, however, because the authors did not remark on changes in the dark controls.

Numerical data from this study can be used in an order-of-magnitude calculation to determine possible effects from exposure to photoflash. Details of the calculations are presented here as an example of the procedures followed throughout this literature evaluation. The lesser light exposure from the daylight fluorescent lamps used in the experiment was about 1.5×10^5 lm·s/cm^2, with ultraviolet light excluded; this exposure caused a ΔE of about 14. An acceptable maximum ΔE might reasonably be 1–2 units over the display lifetime of an object with this light yellow color, or about 0.1 of the experimental change observed. It is desirable to keep flash exposure down to 0.01 of this. So the maximum visible light energy exposure from flash should be no more than $(0.1) \times (0.01) \times (1.5 \times 10^5$ lm·s/cm$^2) = 150$ lm·s/cm^2 for the spectral output of the fluorescent tubes used in the experiment. A very rough idea of the equivalent number of flashes can be obtained by using the total visible output per flash from the xenon lamps. This is 7×10^{-4} lm·s/cm^2 for the point-and-shoot camera and about 0.01 lm·s/cm^2 for studio strobes. These numbers suggest that roughly 2×10^5 and 1.5×10^4 flashes, respectively, could be allowed over the total display lifetime of the object. The numbers are large enough that it is not necessary to refine the estimate by including the fact that the output of the fluorescent tube in the blue region of the spectrum, where the pigment absorbs more strongly, is somewhat greater than the output of the xenon arc in this region.

> **Conservators' Note:** *It can be concluded that handheld flash cameras with ultraviolet blocking are not likely to pose a significant threat to yellow lead oxide pigment, nor is a reasonable amount of studio flash photography likely to do so. However, the possible nonreciprocal response of lead(II) oxide during the time course of the experiment suggests that it might be profitable to further investigate the effect of visible light on this material.*

Lead chromate

The sensitivity of unstabilized lead chromate (chrome yellow) to near-ultraviolet and visible light has long been known (Kühn and Curran, 1986). Exposure to light causes a darkening and/or change in hue. Bacci and coworkers (1999), while investigating the use of this colorant in a tempera formulation as a dosimeter, found that it is a very sensitive indicator of exposure to visible light or elevated temperatures. Approximately two weeks of exposure to daylight fluorescent lights in a light box, with ultraviolet wavelengths blocked by filters, caused an unacceptable darkening of the tempera. Because the effect of light exposure was not completely separated from the effect of elevated temperature, the data cannot be used quantitatively to determine flash exposures. However, the results of this research suggest that the light sensitivity of unstabilized lead chromate in tempera may be similar to that of alizarin lake (see "Natural red and yellow Western colorants," pages 40–42).

Prussian blues

Kirby (1993) summarized information on the manufacture and stability of Prussian blue. Historically, the composition of this pigment appears to have been rather variable and its durability to have varied as a result. The modern pigment is a purer material consisting of ferric ferrocyanide, which may also contain traces of potassium salts. It has been rated moderately durable when tested by exposure to five hundred hours of summer sunlight, or durable when exposed to six months of 10^4 lux radiation from daylight fluorescent tubes.

> **Conservators' Note:** *Clearly, modern Prussian blues should not be affected by reasonable exposure to photoflash sources. Kirby continues to investigate the composition and durability of historic samples of this pigment. Results of these studies, which are expected to be published in the future, are likely to contain information useful to the current topic.*

Copper greens

In an extensive study of copper greens, Schweizer and Mühlethaler (1968) distinguished among several different compounds with regard to light and heat stability as well as to changes in physical forms and appearance upon aging. They showed that malachite, $CuCO_3 \cdot Cu(OH)_2 \cdot H_2O$; two hydrated forms of basic copper acetate, $Cu(CH_3COO)_2 \cdot Cu(OH)_2 \cdot 5H_2O$ and $Cu(CH_3COO)_2 \cdot [Cu(OH)_2]_3 \cdot 2H_2O$; Cu(II) ortho arsenite; and Cu(II) diarsenite undergo light-induced color changes or fading, whereas other forms of these pigments and Schweinfurt green are not particularly sensitive to light. However, the authors report results only for extremely long exposures (1000 hours of xenon-arc radiation, which was four times that necessary to cause Blue Wool no. 7 to fade), and the color changes were not quantitated. Under these circumstances, it is not possible to predict whether the much shorter exposure times from flashlamps would have a significant effect on the pigments.

White oxides, barium sulfate

The photochemistry of the white pigments zinc oxide, ZnO, and titanium dioxide, TiO_2, has received much experimental attention, partly due to the multiple uses of these compounds. Not only can they serve as white pigments in industrial coatings and artists' paints, but they also function as ultraviolet screening agents, fluorescent brightening agents, white opacifying agents, and—especially for titanium dioxide—catalysts of photochemical reactions. Kühn (1986, 173–74) comments that zinc oxide has less tendency to yellow than other white pigments in oil paints, but he refers to somewhat conflicting results regarding the influence of details of the crystalline structure of the pigment on its photochemical activity. Another white pigment, barium sulfate (which is also known as *baryta* when used as a ground layer on photographic print paper), is reported to be highly stable to light. The effects of these pigments on the stability of resins and other materials are discussed in the sections on pigments in nonfibrous polymers and on fluorescent brightening agents later in this chapter.

Natural organic colorants

Dyes have been obtained from plant and animal sources for millennia, and they have been identified chemically with increasing precision over the past century. Mills and White (1994a) present a summary of the more common, naturally occurring organic colorants found in art and archival objects (see corresponding section of chap. 3 for a description of the chemical structures and classifications of natural dyes). Lakes are insoluble pigments made by precipitation of organic colorants onto an opaque base. The lakes described in this section were prepared with natural dyes and an opacifier that is assumed to be inert.

Egerton and Morgan (1970a) review the physicochemical aspects of dye photochemistry, and van Beek and Heertjes (1966) present an excellent introductory overview of the photochemical reactions possible for major types of natural as well as synthetic dyes. More recently, Feller (1994, 177–85) discusses various aspects of the study of dye degradation, including photoinduced changes. Kramer (1986) has published an extensive review of the theory of photophysical processes and the different types of photochemical reactions that various classes of dyes may undergo. He considers relationships between the lightfastness and other physical properties of certain types of colorants, including interactions of dyes with some fibers, as well as photocatalytic fading in mixtures of dyes. Other than mentioning a few excited-state lifetimes and rates of typical reactions, Kramer does not present any numerical data.

Natural Japanese colorants

An investigation of the rates of fading of colorants in Japanese woodblock prints was conducted by Feller, Curran, and Bailie (1984). One of the most sensitive dyes they studied was the blue anthocyanin, *aigami*. The color of this compound is known to be dependent on its state of aggregation (Goto and Kondo 1991). *Aigami* was faded perceptibly after only two days of exposure to soft-white fluorescent lights, suggesting that 10^3 flashes would probably cause perceptible fading of *aigami* in pristine Japanese prints (see the discussion of this in chap. 3). If photoflash exposure is to be limited to 0.01 of this, only about 10 exposures to xenon studio strobes could be allowed over the exhibit lifetime of the object.

In the same study, the yellow colorant *enju* was also observed to be very light sensitive. The degree of sensitivity appears to depend on which part of the plant is used as the source of the dye, which can be as fugitive as *aigami*. Interestingly, *enju* has been identified as a glycoside of quercetin (that is, like *aigami*, it is also a flavonoid glycoside), which raises the possibility that the color and light sensitivity of *enju* also depends on its state of aggregation.

> **Conservators' Note:** *If total flash exposure is to be kept to 0.01 of that which could cause perceptible change in* aigami *or* enju, *as few as 10 flashes from a studio strobe setup without complete ultraviolet blocking could provide the maximum lifetime exposure. Display of pristine Japanese prints should also be kept to a minimum if they are known to contain these extremely fugitive colorants.*

Natural red and yellow Western colorants

Schweppe and Roosen-Runge (1986) quote Feller as stating that carmine, in a lake formulation as distinct from dyed on fabrics, can be as light sensitive as Blue Wool no. 1. Using the data on the Blue Wool standards suggests that a maximum safe flash exposure for this colorant may be on the order of 10^3 flashes.

Saunders and Kirby (1994a, 1994b, 1996) investigated the wavelength-dependent fading of, and the applicability of the reciprocity principle to, various natural colorants that have been used in watercolors. In the first of these studies, they observed the rate of color change when lakes of natural dyes were exposed to daylight fluorescent lights filtered by each of seven broad band-pass filters that together spanned the visible spectrum. Colorants considered fugitive were tested. As expected, they were faded most efficiently by wavelength regions coinciding with their absorption bands. The most sensitive colorant investigated was litmus (the authors' choice to represent fugitive orchil colorants from lichens).

> **Conservators' Note:** *From the extensive data presented by Saunders and Kirby on yellow buckthorn lake, and on the relative sensitivities for the other colorants, it can be determined that orchil or litmus, when used as watercolors, should be limited to a maximum of 10^3 flashes from a studio strobe. Brazilwood lake appeared to be a few times less sensitive, and yellow buckthorn lake about an order of magnitude less sensitive (see chap. 3 for more details).*

Saunders (1995) compared exposure to photographic flash with exposure to gallery lighting of the same total light energy and again observed litmus to be the most light-sensitive of the colorants studied. Lakes of buckthorn, cochineal, brazilwood, and kermes were about one-quarter as sensitive. The less sensitive colorants used in these investigations are listed in chapter 3. The data also indicated a possible lack of reciprocity in the response of some of the lake formulations and of litmus. Buckthorn lake appeared to be more sensitive to photoflash, whereas kermes lake, brazilwood lake, and litmus were somewhat more sensitive to the equivalent total energy delivered by gallery lighting. Agents other than light could contribute to the difference in the latter cases because of the long times required to deliver the total dose.

> **Conservators' Note:** *Of equal concern is the effect of ultraviolet photography on oil paintings on canvas that have these lake pigments. In one second, a 100 W mercury-arc lamp can radiate about 0.01 of the dose of near-ultraviolet and blue photons that caused a perceptible change in the experiments by Saunders and Kirby. Thus, the exposure of an easel painting to a few minutes of radiation from a mercury-arc lamp for the purpose of ultraviolet photography might provide enough radiation to alter the appearance of lakes made from the most sensitive organic colorants.*

Litmus and the buckthorn and brazilwood lakes were also included in Saunders and Kirby's 1996 reciprocity study. (The relevant experimental details are described in the section on lead[II] oxide, page 37). Of the colorants used, only litmus showed major changes at the lowest exposure times; it was altered more than the lead(II) oxide and poor-quality paper. Applying a calculation similar to the one given there indicates that about 10^4 flashes would be enough to cause a perceptible change in litmus and that flash exposures of objects in mint condition must be kept well below this.

Johnston-Feller and coworkers (1984) studied the lightfastness of alizarin lake in an Atlas Fade-Ometer equipped with a xenon arc. Approximate calculations based on their results suggest that this pigment may have a light sensitivity similar to those of the moderately sensitive lakes investigated by Saunders and Kirby (1994a, 1996). (See chap. 3 for an evaluation of the data and a list of the less sensitive colorants reported by these authors.)

> **Conservators' Note:** *Some restriction of photography may be desirable for objects containing alizarin lake. Data from the work by Johnston-Feller's group (see chap. 3) suggests that about 10^3 studio strobe flashes is the maximum allowable for a previously undisplayed object. At least an order of magnitude more flashes from point-and-shoot cameras could be permitted. More flash exposures could actually be permitted if an object containing alizarin lake had already been displayed extensively and had faded somewhat, because the rate of photodegradation decreases with time of exposure.*

Indigo

The natural blue colorants on museum objects are often indigo or indigoids. The stability of indigo in solution has been the topic of a few studies because this dye is sometimes quantitated by extraction and evaluation of the extract. Kuramoto and Kitao (1979), while investigating the mechanism of breakdown of the dye, obtained data suggesting that light equivalent to about 10^5 flashes from a studio xenon strobe with ultraviolet blocked would be required to oxidize a significant percentage of the dye. Green and Daniels (1993) have shown that degradation is greatly enhanced by impurities in the solvent and that light further accelerates this process. When quantitating indigo in solvent extracts, it is advisable to distill the solvents used and to keep the dye solutions in the dark as much as possible.

Indian yellow

Indian yellow illustrates several types of problems that can arise when one attempts to quantitate the effects of photoflash sources. The colorant was prepared from the dried urine of cows fed a diet of mango leaves. This practice has been banned for almost a century on humane grounds; cows could survive no more than a few years on this inadequate diet. The major colored compounds in Indian yellows are salts of the anthraquinone glycoside euxanthic acid (Mills and White 1994a). The pigment was prepared with differing attention to purification of the raw material. Consequently, the lightfastness of Indian yellow can vary by more than an order of magnitude, depending on the sample and the nature of any binders or fillers present (Baer et al. 1986), as well as on the history of the object. Without testing individual samples, a conservator could not predict whether an object containing Indian yellow would be moderately lightfast or much more susceptible to degradation by light. Thus, the risk involved in allowing extensive photoflash exposure cannot easily be predicted with any degree of confidence. Furthermore, the available experimental data suggest that some formulations of this colorant may not obey the reciprocity principle (Baer et al. 1986).

> **Conservators' Note:** *Because Indian yellow is no longer manufactured, research on carefully prepared, well-characterized samples—which might clarify the observed variations in lightfastness—cannot be conducted. Given this situation, a collections manager might prefer to take a cautious approach to unnecessary exposure for materials containing this colorant.*

Green dyes, porphyrins, carotenoids

Porphyrins and carotenoids may be present in natural colorants prepared from plant leaves. For example, sap green is based on chlorophyll. There is extensive literature on the photochemistry of these types of compounds but little research from the perspective of artists' materials. Two recent reports in the photochemical literature (Móger, Köhler, and Getoff 1996; Mortensen and Skibsted 1996) indicate that the quantum yields for photochemical change of these types of substances can be substantially higher (0.001–0.005) than the yields assumed in the example calculations in chapter 1. The yields depend strongly on the environment and state of aggregation of the dye molecules (see chap. 3 for more details). The results of Saunders and Kirby (1994a, 1996), which show that sap green on paper has a light sensitivity similar to that of brazilwood lake, are consistent with the previously mentioned research. These findings suggest that works containing sap green dye on paper should be limited to a lifetime photoflash exposure of about 5×10^3 flashes.

> **Conservators' Note:** *This brief survey of natural dyes used as watercolors or in lake pigments emphasizes information on those colorants that have been shown to be very light sensitive. Some other natural colorants studied were found to be less fugitive; however, absence of a colorant from this survey may also be due to a lack of published data.*

Synthetic organic colorants

An early extensive review of the lightfastness of synthetic dyes was published by Giles and McKay (1963). In their report, they remark on the dearth of quantitative data of the sort required for the purpose of this review. The situation has changed somewhat, although useful numbers must frequently be extrapolated from results in reports that have emphasized other aspects of dye photochemistry. Some recent reviews that provide helpful background information and references are those of Zollinger (1987a), Allen (1987, 1989), Allen and Edge (1992c), and Feller (1994, 177–85). A large proportion of research on the lightfastness of synthetic dyes has been conducted on fiber samples dyed with the colorants. Some results of these studies are discussed later (see "Polymeric Materials Containing Colorants, pages 78–102"); data obtained from experiments on colorants in solution are evaluated here. These investigations have frequently been performed as model studies for the permanence of colorants associated with polymeric materials.

Synthetic organic colorants can be classified according to the nature of the dyeing process and the interaction with substrates, or according to particular chemical features. The former system has led to Colour Index (C.I.) names such as "C.I. Disperse Yellow *x*" and "C.I. Basic Blue *y*." The system based on chemical classification is used here instead. Some major categories of colorants, based on the chemical structures of the dye molecules, are anthraquinones, aromatic monoazo dyes and diazo dyes, flavonoids, azines and oxazines and thiazines, and triphenylmethane dyes. The core structures on which these groupings are based are shown in figure 2.1. Thousands of variations within these groups of chemicals are possible by substitutions and rearrangements of the substituents, which are atoms or atomic groupings chemically bonded to the core structures. Differently colored dyes exist within each group, depending on the substituents. Several authors have remarked that a systematic relationship between the nature, number, and location of

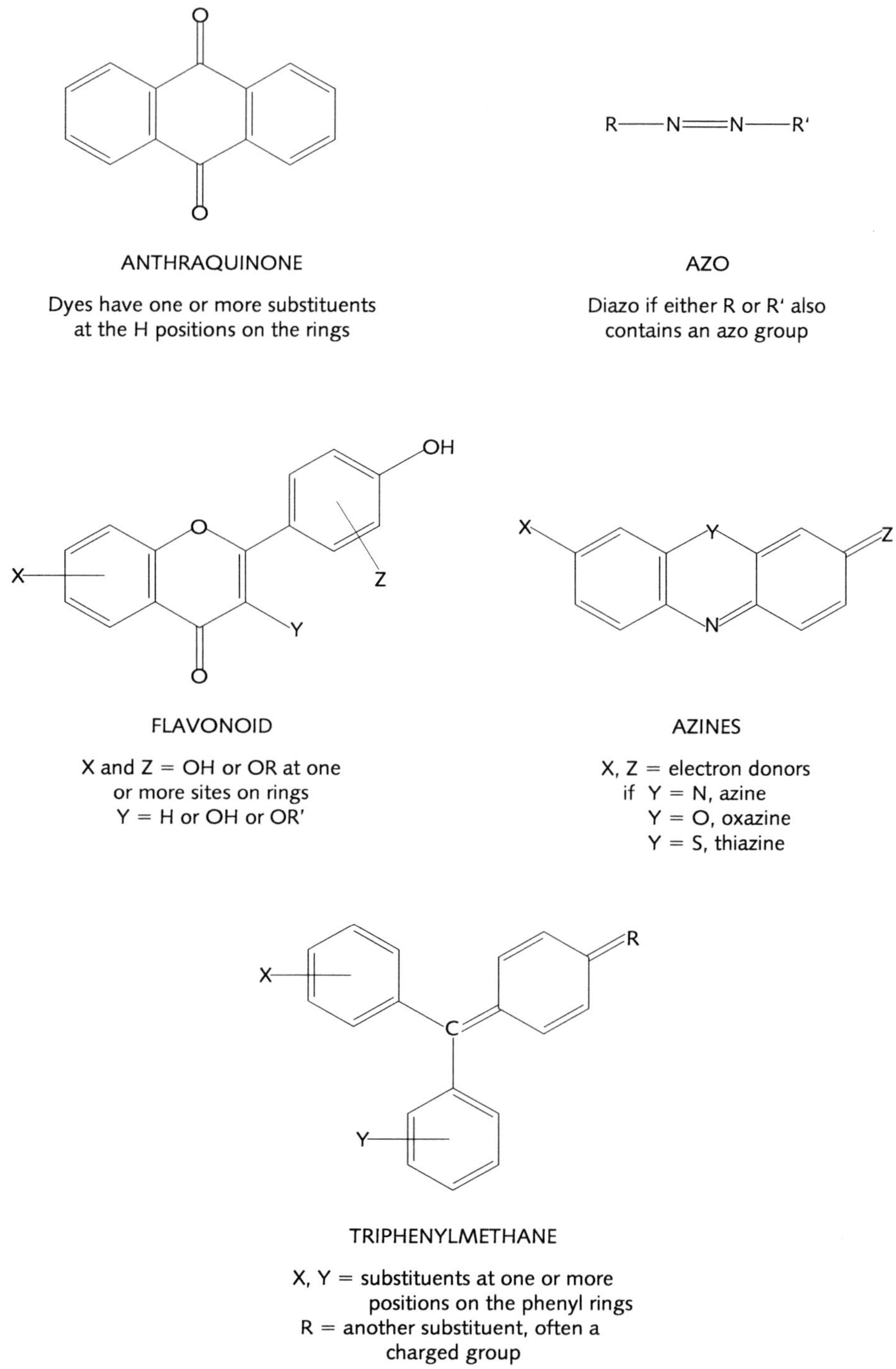

Figure 2.1
Core structures of several different chemical classes of colorants.

substituents and the lightfastness of colorants is not apparent (Giles and McKay 1963; Egerton and Morgan 1970b; Allen and Edge 1992c). The importance of the role of the solvent or substrate, of other components of the system, and of physical conditions has been stressed by researchers (see, for example, Feller 1994, 177–85).

Some early reviews of the photochemistry of synthetic colorants, such as van Beek and Heertjes (1966), provide excellent insight into the chemical reactions that are or may be involved in fading, but they do not supply kinetic data that are relevant here. A great deal of research has been done in an attempt to determine whether the fading of particular classes of dyes and organic pigments is caused by oxidative or reductive processes (see, for example, Griffiths 1980; Allen 1989). One or both of these types of reactions may occur, depending not only on the type of colorant and its detailed chemical structure but also on the chemical nature of the substrate, other substances that may be present, and the availability of oxygen. In the course of investigating this problem, several researchers have published data useful for the present purpose. Theoretical studies have also been undertaken to relate the fading of synthetic organic colorants to their chemical structures, that is, to the distribution of electrons in the molecules. Some data relevant to flash exposures can be obtained from reports of this type of work. Many kinetic studies of the photodegradation of synthetic organic colorants have been performed using ultraviolet light sources emitting wavelengths below the range of interest for the current topic. Occasionally, however, some relevant information can be obtained from such reports. The implications for photoflash exposures of the results of some of these types of investigations are given in the following sections.

Anthraquinone dyes

The lightfastness of a series of amino-substituted anthraquinone dyes has been shown to be directly related to the ease with which reagents that abstract electrons could attack amino- or oxygen-containing groups (Ray and Deheri 1995). The 2-amino and 2,3-diamino anthraquinone dyes in the series tested by Ray and Deheri were more vulnerable than the other amino-substituted anthraquinones by more than an order of magnitude. The difficulty that conservators have with applying this sort of information is that they need to know precisely the chemical structures of the dye or dyes present in an object—information that may not be available without undertaking detailed chemical analyses.

Allen and coworkers (1981) studied the flash photolysis of aqueous solutions of two commercial anthraquinone dyes that contained amino, anilino, and sulfonic acid substituents. The light source was a high-powered xenon arc. Quantum yields as high as 0.1 may have been observed when the aqueous solutions contained *n*-propanol or sodium hydroxide (see chap. 3 for details). The alcohol solvent was used to mimic possible dye surroundings in fibers such as cellulose and rayon that contain many hydroxyl groups. Sensitizers such as benzophenone, which absorbs near-ultraviolet light, markedly increased the amount of fading, while various stabilizers protected the dyes from fading to some extent. An *N*-oxy piperidine derivative was the most effective stabilizer tested.

> **Conservators' Note:** *The data are not adequate to assess quantitatively the effects of xenon photoflash lamps on materials dyed with these particular anthraquinone dyes. They do indicate, however, that situations should be avoided in which artifacts containing these dyes are exposed to near-ultraviolet light with wavelengths of 320 nm or above.*

Kuramoto and Kitao (1981) investigated the role of singlet oxygen in the photofading of several related aminoanthraquinone dyes. The data suggest that, of the colorants they studied, the one most sensitive to visible wavelengths may have a quantum yield significantly greater than 10^{-5} for wavelengths above 510 nm. In the same report, Kuramoto and Kitao indicate that two of the blue anthraquinone compounds they investigated were able to photosensitize the fading of yellow dyes in solution. In these cases, light of wavelengths greater than 510 nm was capable of fading dyes with absorption bands between 418 and 478 nm in the presence of the blue dyes, which absorbed the lower-energy, longer-wavelength light. Unfortunately, the experimental details provided are not sufficient to permit calculation of flash exposures for these compounds.

The sensitized fading of two anthraquinone dyes, a diphenylmethine dye and an indigosulfonic acid dye in organic solvents, was studied by Oda and Kitao (1985). The solutions of sensitizer (rose bengal) and dye were again excited by wavelengths above 510 nm. The data and experimental details suggest the quantum yield for fading was on the order of 0.1 (see chap. 3 for more information). Nickel salts of the dyes or of toluenesulfonic acid appeared to partially protect against photodegradation. Fading of the dyes adsorbed onto silica gel was also investigated; in most cases fading decreased only slightly compared to the extent of fading in solution.

Conservators' Note: *These results suggest that the particular dyes investigated are prone to sensitized photofading in visible light. When these dye mixtures are present in art objects, the objects may have relatively short display lifetimes. Because about 10^3 flashes of a xenon photoflash lamp might cause a measurable amount of fading, excess exposure of these particular dyes to unrestricted photography should probably be avoided.*

Rembold and Kramer (1980) also investigated sensitized fading in anthraquinoid dye mixtures, using monochromatic visible light from dye lasers or a filtered mercury arc. In this study, dimethylanthracene oxidation, which results in fading, was sensitized by a variety of diaminoanthraquinones. The location of the amino substituents on the anthraquinone strongly influenced the fading. Their report is mostly concerned with the development of mathematics to express the kinetics and mechanisms of the reactions leading to fading, but the authors do present some tabulated yield data. Triplet formation, which can lead to photochemical degradation, occurred in the most sensitive mixtures of dyes with a quantum yield greater than 0.5, whereas the less sensitive systems had yields of about 0.01. The most sensitive compounds all had 1,5- or 1,8- substituents.

Conservators' Note: *These data suggest that particular mixtures of dyes that include anthraquinones can be very sensitive to visible light. Objects containing these mix-*

tures would have relatively short display lifetimes and would be able to tolerate fewer flashes before damage became evident than if the dyes were present individually.

Anthraquinone disperse dyes in aqueous acetone were investigated by Giles, Hojiwala, and Shah (1972) as part of an extensive study of a variety of dyes in polymer environments. The quantum efficiencies for fading in solution were generally about one to two orders of magnitude higher than those in polymer film matrices (for data, see under "Polymeric Materials Containing Colorants," pages 78–102).

Triphenylmethane dyes

Duxbury (1993) recently published an exhaustive review of the earlier literature on the photochemistry and photophysics of triphenylmethane dyes, emphasizing structures and mechanisms of the photochemical reactions that have been observed. One short section is devoted to a summary of the effect of the spectral distribution of the light source. The author points out that in some cases the dyes that are less lightfast are faded by visible as well as ultraviolet light, and the ones that are more lightfast are faded by higher-energy, near-ultraviolet light only, as suggested by early researchers such as Egerton (1948) and Giles and McKay (1963). However, this rule of thumb is not followed by other triphenylmethane dyes. Duxbury (1993, ref. 576) also cites the results of Nakamura and Hida, who found that raising the wavelength of light from 250 nm to 577 nm in the solvents 2-propanol and acetonitrile reduced the quantum yield for photodegradation by a factor of 10.

Allen, McKellar, and Mohajerani (1980) also investigated the lightfastness of several basic triphenylmethane dyes in propanol or acetonitrile solvents, which were chosen as models for cotton or acrylic fibers, respectively. They demonstrated that in propanol only, exposure for several hours to light of wavelengths greater than 500 nm from a Microscal light source, followed by 30 minutes of exposure to near-ultraviolet and blue light (wavelengths greater than 350 nm), caused rapid permanent fading of the dyes in the absence of oxygen. These observations are due to the following photochemistry: (1) absorption of the longer-wavelength visible light creates relatively long-lived triphenylmethyl radicals that absorb near-ultraviolet and blue light in the 350–450 nm region; (2) when the radicals absorb light they undergo photochemical decomposition, which results in permanent loss of color. From the information provided by the authors, it can be deduced that the quantum yield for formation of the radicals is small, about 10^{-5}, but the yield for further reaction of the radicals may be about three orders of magnitude higher. In other words, if a buildup of the intermediate radicals has occurred during long exposure to high levels of visible light without oxygen readily available, a sudden exposure to near-ultraviolet light may quickly cause permanent fading.

Conservators' Note: *It cannot be predicted whether the series of reactions observed in this research will occur to a* *significant extent when objects containing these triphenylmethane dyes in the solid state are exposed to photoflash. Because oxygen inhibits radical formation, and because museum lighting levels are usually relatively low, the fraction of dye in the radical form may be very small at any given time in a particular object. Allen's group has reported significant changes in these dyes when they are in films and fabrics, however, and when more intense light containing ultraviolet wavelengths was used.*

The photobleaching of crystal violet and some related triphenylmethane dyes in solution was studied by Kuramoto and Kitao (1982). More recently, the fading of these dyes in acetonitrile was studied by Naguib and coworkers (1996). Crystal violet itself is not very sensitive to visible light because the excited molecule returns extremely rapidly to the ground state, significantly decreasing the possibility for photochemistry. In the presence of appropriate sensitizers, however—which become reduced when they absorb light and then, in turn, donate an electron to the dye to create a radical—crystal violet may be indirectly photobleached. Kuramoto and Kitao observed bleaching with a quantum yield on the order of 0.01 or higher when methylene blue was used as a sensitizer. Naguib's group reports yields as high as 0.2 using benzophenone sensitizer excited at 366 nm; in the absence of sensitizers, the yields are three or more orders of magnitude lower for crystal violet.

Jockusch and coworkers (1996) studied bleaching of several triphenylmethane dyes in three-component solutions containing dye, lactone sensitizer, and an amine. They used laser flash photolysis at 355 or 532 nm, or constant irradiation at 365 nm. Their measurements showed that methylene blue is more easily bleached by the sensitizing system than is crystal violet because several side reactions of the reduced intermediate can occur. Unfortunately, the authors do not present quantitative results for visible (532 nm) excitation.

Conservators' Note: *This finding serves as a reminder that some dye mixtures containing light-sensitive components may be particularly at risk from exposure to visible as well as near-ultraviolet light. If all the colorants present in an object are not positively identified, and/or their possible interactions are not known, an assumption of high light-sensitivity might be the safest course. This can also apply to mixtures of colorants with various degrees of lightfastness.*

Azo dyes

A very useful survey of the chemical and physical properties of aromatic azo pigments was published by Berrie and Lomax (1997). They approach this topic from the point of view of the conservator. Albini, Fasani, and Pietra (1982) measured the quantum yields for photodestruction of the substituted azobenzene dye 4-diethylamino-4′-nitroazobenzene in polar (e.g., methanol) and nonpolar (e.g., methyl *t*-butyl ether) solvents when it was irradiated with some ultraviolet wavelengths or with blue light at 434 nm. The yields for 365 and 434 nm light were consistently on the order of 5×10^{-5} or less for all the solvents tested.

> **Conservators' Note:** *These results suggest that visible and near-ultraviolet light from xenon photoflash lamps is unlikely to affect this particular azo dye.*

Marcandalli and coworkers (1990), in a study of cis–trans isomerization of four related azo colorants in various solvents, found dye isomerization induced by visible light from a 100 W tungsten lamp to be reversible in the dark and to not lead to any photodegradation. They report that photoreduction, for example, was observed only when high-intensity radiation at shorter wavelengths was used. Their estimated quantum yields for these latter processes were always less than 10^{-3} (for the behavior of these dyes on silk see "Polymeric Materials Containing Colorants," pages 78–102).

> **Conservators' Note:** *Damage from moderate exposure to photoflash lamps would seem to be unlikely for these four dyes, provided that the ultraviolet light is blocked.*

Sirbiladze and coworkers (1990) investigated the lightfastness of some azo dyes in terms of the relation between the kinetics of the photoreactions and the formation of radicals. A series of structurally related dyes in a 2-propanol–water (10:1) solution was tested in the presence of a radical-forming initiator. The results suggest that the quantum yields for solution photochemistry of the more sensitive dyes, when irradiated with the 365 nm band from a mercury-vapor lamp, were about 0.1.

The sensitized photolysis of several red, orange, or violet acidic azo dyes was studied in aerobic aqueous solutions by Neevel, van Beek, and van de Graaf (1992). The light source consisted of two xenon flashtubes on either side of a tubular cuvette; a filtering solution surrounded the cuvette so that only near-ultraviolet light absorbed by the sensitizer reached the dye solution. Although the dyes were structurally somewhat similar, the presence of various substituents in different locations caused the photochemistry to proceed by three different reaction mechanisms, as suggested by the transient absorption spectra observed following flash photolysis. Also, the only dye tested that did not form a hydrazone in solution—C.I. Acid Red 37—remained measurably faded after only one flash in the presence of an excess of sensitizer. The numerical information provided in the report is incomplete but allows a guesstimate for a quantum yield for this dye on the order of 0.001 or higher. The authors state that longer irradiation times resulted in permanent fading of C.I. Acid Red 37, but the relevant data are unpublished.

Other synthetic dyes

Nansheng and coworkers (1997) compared the effects of sunlight and artificial ultraviolet light on the fading of two basic dyes (Basic Yellow 2, a ketone imine dye; and Basic Red 13, a substituted indole derivative of a methine dye) in aqueous solution in the presence of ferric (Fe^{3+}) salts. They observed significantly faster kinetics for sunlight. Some of the observed increased rate of fading may have been due to absorption by the dyes of the visible and near-ultraviolet wavelengths not provided by the artificial ultraviolet light. These wavelengths are also emitted by xenon flashlamps. In contrast, three reactive dyes (C.I. Reactive Red 2, C.I. Reactive Blue 4, and C.I. Reactive Black 8) were not as light sensitive.

> **Conservators' Note:** *Combining the data of Nansheng's group with reasonable assumptions regarding the spectrum of sunlight compared to the output from a xenon flashtube suggests that these moderately sensitive dye samples might be significantly affected by about 10^5 flashes of a xenon photostrobe in the presence of ferric ions. This is not surprising, because Fe^{3+} is known to be a good sensitizer of oxidation processes in organic chemistry.*

While testing the efficacy of a newly developed microfading apparatus for determining the stability of colored museum objects, Whitmore, Pan, and Bailie (1999) exposed several gouache paints to constant radiation from a xenon-arc lamp, with ultraviolet and infrared wavelengths removed by filters. Rose tyrien, rose bengal, and magenta were the most sensitive paints studied. In less than one minute they faded to an extent that would be detected by eye if the fading were to occur over a large enough area. If the xenon light source in the microfading apparatus has the same spectral output as camera flashlamps, about 2×10^5 flashes would be the maximum allowable over the display lifetime of an object containing rose tyrien gouache. Rose bengal and magenta gouaches could endure about 3×10^5 flashes from a handheld camera (see chap. 3 for numerical details).

> **Conservators' Note:** *Provided that ultraviolet light is removed from the flash, a moderate amount of gallery photography of art objects containing the gouache paints rose tyrien, rose bengal, or magenta may be permissible.*

Matsushima, Mizuno, and Itoh (1995) investigated the solution photochemistry of substituted 2-hydroxychalcones, which are model compounds for anthocyanins. These substances are converted to flavylium ions by blue light when in aqueous ethanolic solutions, and they revert to the substituted chalcones when heated to 50 °C in the dark. The flavylium ions have strong absorption bands in the 520–560 nm region, which is why they appear blue, whereas the chalcones absorb near-ultraviolet and blue light between 380 and 410 nm. The researchers used the 406 nm band from a mercury arc to irradiate the solutions. They found that the cyclic photochromic behavior of the colorants in solution could be repeated at least ten times without any indication of fatigue, even when the solutions were irradiated with more than 10^{21} photons. When the substituted chalcone model compounds were included in polymer films, however, the flavylium ions tended to be bleached by repeated cycling (see the discussion in the section on dyed polymeric materials).

Photobleaching of the substituted azine dye phenosafranine (C.I. 50200) is sensitized by a lactone in the same way as is methylene blue (Jockusch et al. 1996). The dye is also less lightfast than crystal violet (see earlier discussion).

This sampling of the literature on the photodegradation of synthetic colorants in solution indicates that these materials display a very wide range of sensitivities to light. Although in the majority of cases the quantum yields for dye photochemistry are significantly lower in the solid state than in solution, this is not always the case. Sensitization of one colorant by another in mixtures may occur, but it is not necessarily predictable on the basis of known structures or absorption spectra of the components. Solution studies have provided considerable insight into the mechanisms of photodegradation of various classes of synthetic colorants.

Natural Fibers

In a general review of textile conservation, Crighton (1992) has presented a useful qualitative discussion of the physical aspects of fiber degradation by light, with an emphasis on natural fibers.

Cotton

Textile researchers have been investigating the effects of light exposure on cotton yarns or fabrics for several decades. These efforts are summarized by Buschle-Diller and Zeronian (1993b) and by Rånby (1989). Most of the earlier results do not provide numerical data useful for the current purpose. Pure cellulose, regardless of the source, should not absorb light of wavelengths greater than 310 nm (Phillips and Arthur 1985). Ultraviolet light of shorter wavelengths may lead to chain scission, and wavelengths less than 200 nm can cause further breakdown of the cellulose with the release of small, volatile molecules. In the presence of oxygen, various oxidation products may be obtained. If cellulose has already been oxidized or yellowed due to the effects of heat or chemicals, however, it is likely to absorb near-ultraviolet and blue light (Tylli, Forsskåhl, and Olkkonen 1993). Responses to light exposure of yellowed or oxidized cellulose in the form of fibers or sheets have been studied by several investigators. Most reports provide incomplete information on the light sources used, but some contain enough details to make approximate calculations possible. There is evidence that near-ultraviolet light can cause yellowing and that blue light may induce bleaching of cellulose yellowed by heat (Silvy and Le Nest 1976), resulting in an apparent photochromic effect.

Conservators' Note: *The similarity between the latter phenomenon and aqueous light bleaching of discolored paper serves as a reminder that the amount of light energy provided by a photographic flash without ultraviolet light need not have a deleterious effect on cellulose.*

The effect of 350 nm irradiation on celluloses has been investigated by Reine and Arthur (1970) for purified wood cellulose, and by Tylli, Forsskåhl, and Olkkonen (1993) for various purified α- and β-celluloses. The latter researchers, using the exciting light source in a spectrofluorometer to irradiate their samples, detected slight changes in fluorescence after 15 minutes of irradiation. The extent of change increased as the time of irradiation increased. Reine and Arthur reported a 10% decrease in fiber strength after 2 hours of radiation from a source that emitted about 5×10^{16} photons/s at 350 nm. The emission from these sources at 350 nm is similar to that from the mercury arcs used for ultraviolet photography.

Conservators' Note: *The results of these studies suggest that ultraviolet photography requiring several minutes of exposure might damage cotton fabric or cellulose papers. The benefit of ultraviolet photography to conservation should be weighed carefully against the possible deleterious effects of the procedure. Conversely, these results suggest that xenon flashlamps that irradiate the cellulosic object with light of wavelengths greater than 360 nm should not pose a significant threat.*

Using a weathering device with a xenon arc filtered to simulate sunlight through window glass, Jeon and Block (1990) demonstrated that raising the temperature, the relative humidity, or both will increase the rate of photodegradation of a washed cotton test fabric. The textile studied appeared to be about an order of magnitude more sensitive to increased relative humidity than to elevated temperatures at a given relative humidity. Under conditions that might be considered extreme for most museum environments, however—about 30 °C and about 75% RH—the data suggest that roughly 50 hours of exposure to the xenon arc would be required to cause a significant decrease in fabric strength.

Conservators' Note: *It can be concluded from this study that moderate exposure to xenon photoflash lamps, which emit several orders of magnitude less near-ultraviolet light than the light source in a weathering device, would not be deleterious to the fabric even under the less-than-ideal environmental conditions used in the investigation.*

Buschle-Diller and Zeronian (1993a) studied the effect of irradiating cotton cellulose yarns with a xenon arc in a weathering device. The samples had been chemically oxidized to different extents by procedures chosen to create dialdehyde (carbonyl), dialcohol, or dicarboxylic acid derivatives in the cellulose. These oxidized yarns initially lost tensile strength somewhat more rapidly upon irradiation than did unmodified control yarns and re-reduced yarns. After several hundred hours of irradiation, however, the rate of change in strength of the modified and control yarns as a function of time of light exposure appears to have been about the same. The observed changes in strength coincided with the appearance of breaks in the fibers as seen with scanning electron microscopy. No change in the color of the yarns was noticed. The amount of radiation used in this study is rather large, however. Well over 10^8 flashes from a handheld camera would most likely be needed to cause similar changes in oxidized, undyed cotton cellulose, suggesting that on the order of 10^6 flashes over the display lifetime of the object could be allowed (see chap. 3 for the numerical details).

> **Conservators' Note:** *Oxidized cotton cellulose should not be at high risk for photodegradation from exposure to flashlamps emitting wavelengths above 320 nm.*

Silk

Reagan and coworkers (1989) used undyed silk as a control fabric for investigation of the effect of fluorescent whitening agents on dyed silks (see "Fluorescent Whitening Agents," pages 102–8). They reported that their controls—undyed silk samples—had yellowed significantly after receiving 40 American Association of Textile Chemists and Colorists (AATCC) fading units of light from a fading device. This is equivalent to the amount of light needed to cause a color change in Blue Wool no. 5—about 3×10^7 lux hours (Colby 1992). The dose of light used in this study is many orders of magnitude greater than that emitted by camera flashlamps.

In a study of the effect of light on the degradation of silk fibers, Kuruppillai, Hersh, and Tucker (1986) evaluated the effect of xenon-arc radiation on silk habutae, a lightweight (3.14 mg/cm^2), plain-weave, unsized, undyed fabric. A measurable change in the breaking strength of yarns and an increase in the number of free amino groups were observed after the fabric had been in the fading apparatus for two days. If it is assumed that this was equivalent to 48 hours, which is more than 1.7×10^5 seconds, comparison with the calculated number of photons emitted from a xenon flashlamp during that time (see table 1.2) suggests that the experimental dose is again many orders of magnitude greater than the dose from flash. Consequently, moderate amounts of flash photography with xenon flashlamps should not pose a threat to this undyed silk fabric.

Becker and Tuross (1992) also used silk habutae as a control fabric to compare yellowing by a known source in the Atlas Weather-Ometer to the discoloration of aged silk gowns that were presumed to have once been white. The exposures used to discolor the control fabric were, like those cited earlier, orders of magnitude greater than moderate amounts of flash photography might be expected to deliver. The same authors also report an approximate 10% loss of the amino acid tyrosine from silk irradiated with a total of 100 kJ/m^2 (equivalent to 10 J/cm^2) of light above 300 nm (Becker and Tuross 1994). Without knowing what fraction of the loss was due to ultraviolet wavelengths below 365 nm, it is difficult to draw a specific conclusion regarding photoflash exposures from these data. However, the fact that tyrosine absorbance above 300 nm is rather low suggests that xenon photoflash exposures would not be a significant cause of photodegradation of undyed silk proteins (see chap. 3 for more details).

An ongoing, extensive investigation of the aging of weighted silks is summarized by van Oosten (1994). She comments that light had affected the degradation of historic samples. However, light was not used in the artificial thermal aging of the experimental samples studied up to that point.

Kelly, Mollah, and Wilkinson (1990) report results of diffuse reflectance laser flash photolysis of a commercial silk fabric identified as "raw silk." A rather short-lived, transient absorption band was observed in the visible region after this flash. No fluorescence was detected, whether or not oxygen was present. Presumably this irradiation led to a chemical reaction in the fabric that did not involve an oxidation. To determine whether xenon photoflash could cause a significant amount of this reaction, the time element must be considered. Because of the small number of photons per square centimeter of wavelengths below 360 nm emitted by a xenon flashlamp during a flash, which is very long compared to that of the laser flash, and the large number of amino acid residues per square centimeter of fabric (roughly 10^{19}; see chap. 1), a very small portion of the monomer units of the silk can be excited long enough to react chemically before the excitation energy is dissipated (see chap. 3 for more details).

> **Conservators' Note:** *The probability of photochemical reaction in undyed silk appears to be low. It may be concluded from the work of the investigators cited here that undyed, unweighted silk fiber or fabric is not at high risk for degradation from moderate exposure to photoflash. On the order of 10^5 flashes from a handheld camera without near-ultraviolet blocking or 10^4 exposures from a studio strobe setup could be allowed without undue risk to the integrity of the material over its display lifetime.*

Wool

Lewis (1989) reviewed the evidence for light-induced bleaching and yellowing of wool fibers. The most effective wavelengths for photoyellowing are below 300 nm. Light of 320 nm is 6–7% as effective as more energetic ultraviolet light, and light in the near ultraviolet ($\lambda > 320$ nm) is only very slightly active in yellowing. However, visible light does appear to be involved in the bleaching of cream-colored wool. No data that can be used to quantitate the effects of photoflash lamps are presented in this or earlier reviews of the subject (e.g., Nicholls 1980). Mansour, Cornell, and Holt (1988) report that the presence of residual chlorinated solvents from cleaning processes will significantly increase the photoyellowing effect, but they do not provide numerical data on which calculations for flash exposure can be based.

Cardamone and Marmer (1994) used an 80-hour exposure in a weathering device to cause major changes in the whiteness and yellowness indices of woven wool flannel. The exposure was equivalent to about 2×10^{22} near-ultraviolet photons/cm^2. To evaluate the possible risk of flash, the minimum exposure that might cause a just-detectable change must be estimated. For the purposes of this approximation, only 0.001 of the experimental dose, or about 2×10^{19} photons/cm^2, is assumed to represent 0.01 of the exposure that could cause detectable color change. Roughly 10^5 flashes from xenon strobe lights would be required to provide this many photons.

> **Conservators' Note:** *Flash photography does not appear to pose a significant danger to undyed woven wool flannel.*

The fluorescence of individual wool fibers was studied in detail by Collins, Davidson, and Hilchenbach (1994). They used a helium–cadmium continuous laser, which emits blue light at 442 nm, to irradiate a 0.2 cm^2 area of fibers on a microscope stage for 390 minutes, and then a mercury-arc lamp (365 nm) or an argon-ion laser (488 nm) to excite fluorescence in exposed and unexposed portions of the irradiated fibers. The helium–cadmium laser exposure (roughly 3×10^{21} photons at 442 nm) significantly reduced the intensity of fiber fluorescence.

> **Conservators' Note:** *The amount of visible laser light used in this study is many orders of magnitude greater than the amount delivered by photographic flashlamps at the same wavelength. Consequently, no limit on flash photography is indicated by this research.*

Feller (1994, 169–76) summarizes the earlier results of Australian researchers who demonstrated that 3 minutes of 365 nm light from a mercury arc could cause a slight change in the reflectance of wool fibers. The dose was roughly 10^{18} photons/cm^2 delivered over a 3-minute period, or about 5×10^{15} photons/cm^2/second. This is less than the exposure that an object being photographed under ultraviolet light would receive. Also, the experimental radiation was only a few hundred times the amount of near-ultraviolet light from a studio strobe setup, if the lamps do not have blue color-correcting filters.

> **Conservators' Note:** *These data suggest that conservators should avoid ultraviolet photography of woolen objects and that limiting photography of undyed woolen objects might be an appropriate precaution if it cannot be ascertained that near-ultraviolet light is removed from the flash source.*

Pulp, Paper, and Wood

The literature on the photooxidation of cellulose as well as papers containing groundwood pulp is reviewed by Havermans and Dufour (1997).

Cellulose and rag papers

The comments on cotton yarns in the previous section are also relevant to cellulose papers. Research on the effects of light on cellulose papers has frequently been conducted with unsized Whatman filter papers; these papers have also been employed as unsized controls in some experiments. One example is the investigation by Lee, Bogaard, and Feller (1994), who compared the effects of light on Whatman no. 42 filter paper discolored by thermal aging to the responses of papers made from groundwood pulp and unbleached pulp. The aged filter paper was slightly bleached by a few days of exposure to a xenon arc in an Atlas Weather-Ometer; to black-light fluorescent lamps; or to daylight fluorescent lamps. In the last case, the amount of bleaching decreased as the ultraviolet and blue wavelengths (i.e., the higher-energy components) were progressively removed from the radiation. Very slight bleaching was still detected after three days, however, even when all light of wavelengths less than 460 nm had been removed by a cutoff filter. If it is assumed that the spectral power distributions of the filtered fluorescent source and a xenon flashlamp are similar, it can be approximated that between 10^7 and 10^8 flashes of the latter might cause slight bleaching of yellowed cellulose paper (see corresponding section in chap. 3). If

Figure 2.2

Plot of data from Whitmore and Bogaard (1994), showing three measures of the degradation of unsized cotton cellulose paper (Whatman no. 42)—occurrence of chain scissions; formation of carbonyls; and degree of polymerization—as a function of exposure time to two different light sources: UV-A and daylight fluorescent.

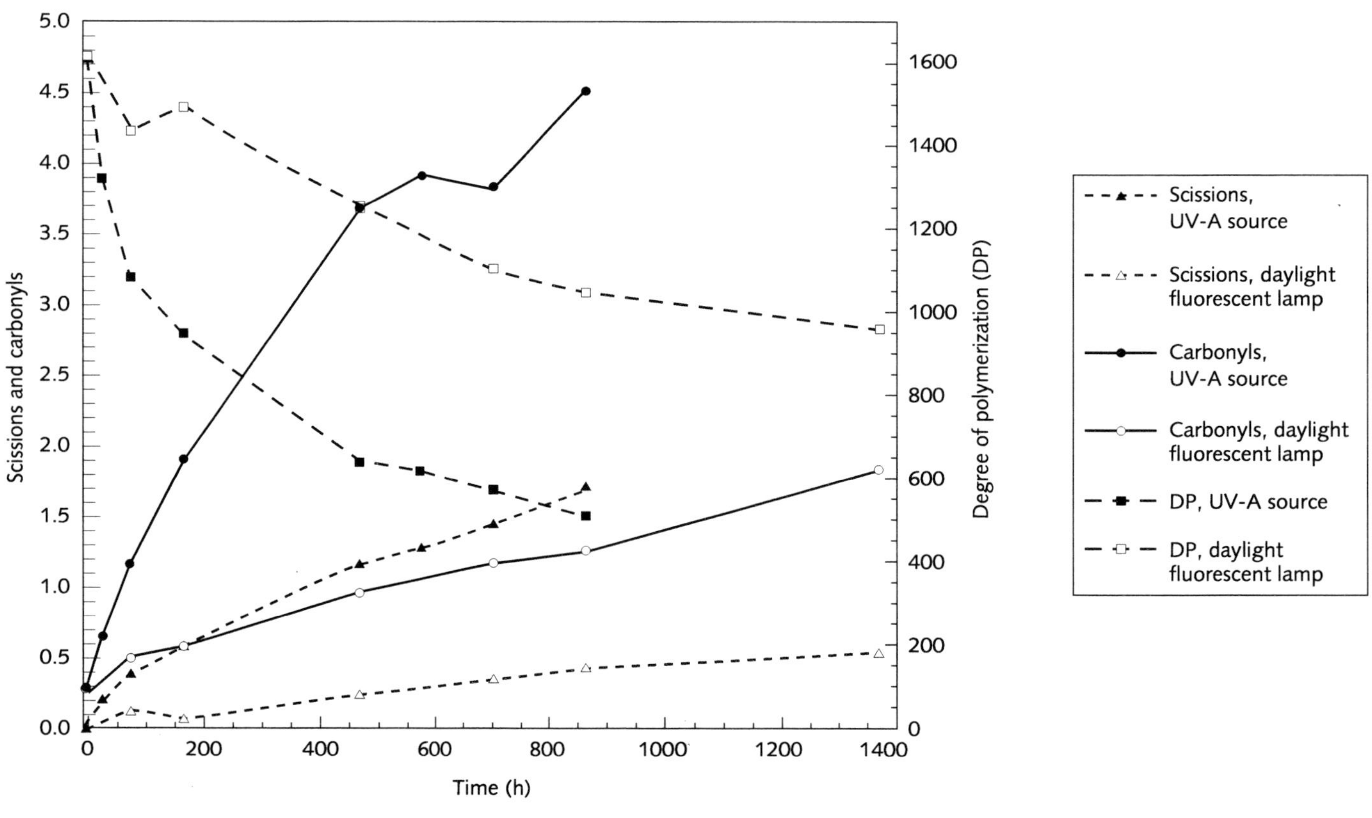

photoflash is to cause no more than 0.01 of a perceptible change, then about 10^6 flashes may be safely allowed.

Lee, Bogaard, and Feller also observed significantly greater bleaching if the paper was immersed in alkaline buffer during exposure, conditions similar to those used for the aqueous light-bleaching treatment of yellowed rag papers. In an earlier study, Feller, Lee, and Bogaard (1986) demonstrated that Whatman no. 42 paper that had been exposed to more than 50 hours of ultraviolet irradiation would degrade more quickly upon subsequent artificial (thermal) aging than paper that had not received the ultraviolet treatment. These results suggest that badly degraded cellulose paper might be more sensitive to additional light exposure than cellulose paper in good condition. The chemical composition of the degraded material may influence the course of further change.

Whitmore and Bogaard (1994) characterized the oxidative degradation of unaged Whatman no. 42 cellulosic paper caused by ultraviolet-radiation or by light from fluorescent lamps. They measured the number of carbonyls added to the cellulose, the number of chain scissions, and the decrease in degree of polymerization of the cellulose chains as a function of exposure. Their report also provides thorough information on the two light sources they used. If their data are reanalyzed as a function of time of exposure (see fig. 2.2), it can be seen that the ultraviolet-A source caused a more rapid deterioration of the paper over time. However, if the data are instead replotted as a function of the total dose of ultraviolet photons/cm^2 (see fig. 2.3), it becomes apparent that the fluorescent lights

Figure 2.3

Plot of data from Whitmore and Bogaard (1994), showing the degradation of cellulose as a function of the total dose of ultraviolet photons received from two different light sources: UV-A and daylight fluorescent.

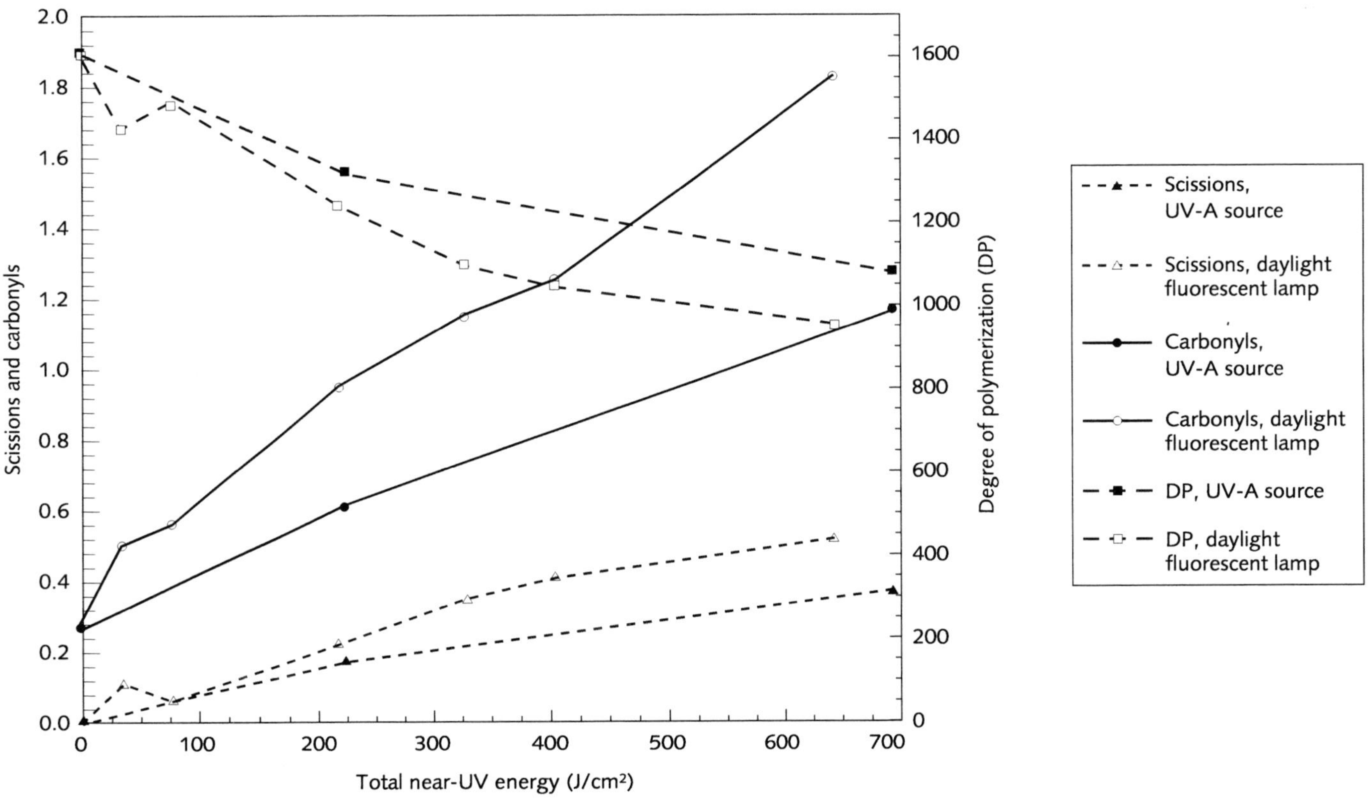

caused the same or greater deterioration per unit of ultraviolet energy delivered to the paper (see corresponding section of chap. 3). The additional deterioration caused by the fluorescent lamps might possibly be due to the visible light they emitted. Further experiments would be required to confirm this.

Evaluation of the data supplied by Whitmore and Bogaard suggests that more than 2×10^7 flashes from a xenon flashlamp would be required to cause the same sort of damage they observed in the samples after three days in the experimental setup (see chap. 3 for details). Thus, about 2×10^5 flashes might be considered a safe total exposure for cellulose papers. For objects glazed with ultraviolet-blocking plastic, or displayed in vitrines constructed of this material, a higher number of flashes could be allowed.

> **Conservators' Note:** *Based on the results of these investigations and others on the effects of light on cellulose papers, it can be concluded that a moderate amount of xenon photoflash exposure should not cause significant damage to these papers in good condition. It should be kept in mind, however, that such experiments have not been performed on badly deteriorated samples (see chap. 3 for specific recommendations on particular papers and light sources).*

Lignin-containing papers

The photoyellowing of papers containing lignin is currently of considerable interest because of the demand for permanent papers, that is, papers that will retain their strength and appearance over a very long time under normal use and storage conditions (McCrady 1998). Investigations have been performed on model compounds for lignin, on pulps, on handsheets made from various kinds of pulps, and on commercial papers. Some of these studies have provided data relevant to the possible effects of flashlamps and photocopier light sources on paper-based art objects and archives.

Heitner summarizes the photochemistry of lignin, which is believed to be involved in yellowing (1993b), and the earlier attempts to lessen this effect (1993a). Lignin is an extensively branched compound. It may be superficially described as a network of phenyl groups that have oxygen-containing substituents such as –OH and $-OCH_3$ on them, connected by short alkyl or vinyl chains. When lignin absorbs ultraviolet light, phenoxy free radicals are formed that can abstract a hydrogen atom from other phenyl groups. These activated groups can in turn cause breaking of bonds and the formation of colored moieties. In the presence of oxygen, peroxides are also formed on the lignin; they will interact with other substituents in the network to form more phenoxy radicals, and so on. Inhibition of these processes has been attempted by reduction of oxygen-containing groups on the lignin, by blocking of other substituents believed to be involved in the yellowing reactions, and by addition of various ultraviolet absorbers and radical scavengers. These attempts have generally produced disappointing results in the long run (e.g., Cole and Sarkanen 1987; Francis et al. 1991).

Research done in the 1990s provides more information on the yellowing phenomenon. The photochemistry of lignin model compounds has been investigated in solution, in films, and with the compounds impregnated into filter paper. Several researchers, using ultraviolet-A light sources with wavelengths less than or equal to 355 nm, observed high quantum yields—up to 0.5—for photochemical changes that involved free radicals (see corresponding section of chap. 3 for specifics and additional references). Apparently, the active wavelengths used in these investigations were always below 360 nm.

Near-ultraviolet laser radiation has also been used by several investigators to perform flash photolysis studies on the yellowing of various pulps (Schmidt et al. 1990; Schmidt et al. 1991; Wilkinson, Goodwin, and Worrall 1993). These experiments indicated that quantum yields could be as high as 0.004 (see chap. 3 for more details).

> **Conservators' Note:** *Results from these studies suggest that although a moderate number of xenon camera flashes should not be deleterious, limiting the extent of exposure of ligneous papers to lamps with a significantly larger near-ultraviolet component, such as photocopier light sources, might be appropriate.*

Tylli, Forsskåhl, and Olkkonen (1995) have investigated the effects of 350 nm radiation from the light source in a Shimadzu spectrofluorometer on the fluorescence and absorption spectra of unbleached, peroxide-bleached, and ozone-bleached groundwood pulps. The ozone-bleached pulp began to show measurable changes within 15 minutes (see chap. 3 for calculations of equivalent flash exposures).

> **Conservators' Note:** *These data do not suggest that moderate flash photography with filtered xenon light sources would be damaging. They do indicate, however, that photocopying should be somewhat restricted. About 10^3 copies would be the maximum allowable over the lifetime of a document on ozone-bleached, groundwood pulp paper.*

Studies on the effects of cyclic irradiation regimens on lignin-containing pulps indicated that light exposure can result in subsequent changes in the dark. Photochromism or synergistic effects (or both) due to subsequent irradiation with visible light may occur. Ek and coworkers (1993) observed that the photoyellowing of bleached or unbleached groundwood pulp caused by exposure to near-ultraviolet plus visible light could be partially reversed by dark storage and that the paper was yellowed further by additional simultaneous exposure to ultraviolet and visible light. The cycle could be repeated several times, with the pulp becoming more yellowed after each experimental exposure, as if it had been constantly exposed to the combined source. Alteration of the effects in the presence of reducing agents or radical scavengers led the authors to suggest that the yellowing involved free radicals (see chap. 3 for a discussion of this interpretation).

> **Conservators' Note:** *Unfortunately, the authors do not provide specific information about their light source, so no calculations can be made to determine how much exposure to flash sources might cause this phenomenon. It should be recognized however, that periodic exposure of displayed objects to flash sources could be construed to resemble the light regimen used in these experiments.*

All of these studies on pulps have used light sources with a major ultraviolet component. The results reinforce the point that ultraviolet exposure of lignin-containing papers should be avoided when possible, including restriction of photocopying. These research reports do not provide the type of numerical data that would permit calculation of exposure limitations for xenon photoflash sources with their lower ultraviolet content, however.

Several researchers have investigated the effects of near-ultraviolet and visible light on handsheets prepared from various pulps. For example, Robert and Daneault (1995) determined that minor changes would occur in papers made from a thermomechanical pulp after they received a total irradiation equivalent to about 10^6 xenon flashes without ultraviolet blocking (see chap. 3).

> **Conservators' Note:** *On this basis, moderate amounts of flash photography with filtered xenon sources could probably be allowed, but, again, photocopying might best be restricted.*

Forsskåhl and Tylli (1993) showed that exposure to 2 hours of blue-light radiation will cause a significant yellowing of handsheets made from chemimechanical pulp. This change is accompanied by a decrease in near-ultraviolet reflectance. These data were interpreted with regard to photoflash exposure by comparing the output of the experimental light source to the blue light emitted by xenon studio strobes and built-in camera flashlamps (see chap. 3). The results suggest that chemimechanical pulp is moderately sensitive to blue light.

> **Conservators' Note:** *A few thousand flashes from a studio strobe setup or about 10^5 flashes from a handheld camera may be the maximum allowable for papers made of chemimechanical pulp. Photocopying would be at least an order of magnitude more damaging and should be restricted to about one hundred copies to avoid decreasing the display lifetime of documents on this type of paper.*

Forsskåhl and Maunier (1993) also performed photocycling experiments on handsheets made from bleached and unbleached thermomechanical pulps, using light of particular wavelengths. Certain near-ultraviolet and visible wavelengths caused significant changes in unbleached-pulp paper, whereas blue light caused significant changes in the bleached-pulp paper in some cases. The authors specified sample irradiances, which have been used to calculate permissible flash exposures (see chap. 3).

> **Conservators' Note:** *Evaluation of data from this study suggests that about 10^4 flashes from a studio strobe might be the maximum allowable over the lifetime of papers made of bleached thermomechanical pulp.*

Forsskåhl and Maunier also observed that after a single exposure to more-energetic ultraviolet-B light, which caused yellowing of the paper, the appearance of the samples could be cycled by first exposing them to blue light and then to near-ultraviolet light of wavelength 370 nm. During this treatment, there was a gradual shift of appearance toward increased yellowing. The cycle could be repeated many times before the visible light could no longer decrease the yellowing caused by the ultraviolet exposure. The authors propose two possible chemical explanations for their observations. The first is a photochromic effect involving photooxidations that are not completely reversible so that eventually the yellow compounds predominate. The second is somewhat more complex: Near-ultraviolet light might convert the absorbing groups to yellow products that can sensitize destruction of other components and that are also subject to alteration or destruction by blue light. These reactions would proceed from the surface, going deeper into the paper sheet as each layer of chromophores is degraded. In support of this last suggestion, the authors cite references showing photoyellowing of papers as a surface phenomenon, but they do not decide between the two possible mechanisms.

An important point illustrated by these experiments is that once this type of paper has undergone an initial yellowing due to exposure to higher-energy ultraviolet-B light—unfiltered sunlight, for example—it is subject to progressive deterioration by cycles of near-ultraviolet light (370–400 nm) and blue light (approx. 440 nm). Although the doses used in the experiments are several orders of magnitude greater than those delivered by xenon flashlamps, the photochemical changes in the paper would be expected to proceed on a smaller scale with weaker light sources.

> **Conservators' Note:** *Precautions should probably be taken to ensure that lignin-containing papers that have been exposed to sunlight or to other sources of ultraviolet light are not subsequently subjected to cycling between light sources emitting visible and near-ultraviolet light. For example, repeated photocopying on a machine*

with a light source lacking ultraviolet blocking, interspersed with display of the object, might have this effect. Further experimentation on this phenomenon, using conditions similar to those in a museum setting, should probably be pursued.

Kim and coworkers (1989) used three types of bleached Kraft pulp handsheets—unsized, alum-rosin sized, and alkylketene dimer sized—to study the response to radiation from an ultraviolet lamp with a maximum output at 365 nm. The brightness of the rosin-sized paper initially decreased at about twice the rate as that of the other samples. All the papers were significantly discolored and had lost fold endurance within 2 hours of exposure. These results add to the evidence that exposure to photocopier light sources with significant near-ultraviolet output might wisely be restricted to avoid decreasing the useful lifetime of ligneous—in this case, Kraft—paper artifacts.

The research group led by Feller used unbleached Kraft pulp sheets and groundwood sheets as well as aged filter paper in their experiments on the effects of light on various papers (Lee, Bogaard, and Feller 1989, 1994; see also "Cellulose and rag papers," pages 56–57). Some experimental details from these studies are given in the corresponding section of chapter 3. In the earlier experiments, aged filter paper and handsheets prepared from groundwood pulp were darkened by exposure to the xenon source in an Atlas Weather-Ometer. The Kraft papers were bleached by the same exposure. However, all *acidic* samples darkened when stored in the dark following light exposure. Groundwood paper samples darkened no matter what their pH. In the more recent study, these investigators observed slight bleaching of the Kraft pulp paper by visible light. Near-ultraviolet light had a much greater effect. The groundwood paper, in contrast, was not affected by visible light but was markedly darkened by exposure to near-ultraviolet light. The magnitudes of the responses were again dependent on the pH of the papers, and the rates of change decreased when exposure times exceeded 48 hours.

Conservators' Note: *This last observation suggests that groundwood pulp paper that has already been badly darkened by light exposure may not be significantly further affected by additional short exposures to visible light from a filtered and color-corrected flashlamp.*

In an intriguing study, Oye (1989) determined the effect of 3 hours of monochromatic radiation at various wavelengths in the near-ultraviolet and visible regions on the brightness and folding endurance of book papers. Significant decreases in brightness were observed only for wavelengths less than about 430 nm, with a major decrease caused by 340 nm light. Folding endurance was affected to about the same extent by each of the wavelengths used between 290 and 680 nm. Unfortunately, neither the light source nor the experimental setup was described, so calculations relevant to flashlamp exposures cannot be made.

Andrady and coworkers conducted two studies on the effect that light of various selected wavelengths has on the appearance of newsprint. In the first case (Andrady et al. 1991), fresh newsprint and samples that had been preexposed to white light were yellowed by exposure to monochromatic ultraviolet light at 340 nm or less. The shortest wavelengths appear to have been the most efficient, causing the greatest increases in yellowness index despite the intensity of the source being lower at these wavelengths. In another experiment where roughly equal numbers of photons at the different wavelengths were employed, the fresh newsprint showed very slight changes when irradiated with 400 or 600 nm light, in addition to larger changes due to ultraviolet radiation (see chap. 3 for calculations based on the experimental data).

> **Conservators' Note:** *The authors' results suggest that about 10^3 flashes from a studio strobe would deliver a dose equivalent to the exposure that caused a slight change in the newsprint in this study. Photoflash and photocopier exposure of newsprint should, therefore, be restricted.*

In the second study, Andrady and Searle (1995) used cutoff and band-pass filters to determine the effectiveness of different spectral regions in causing changes in the yellowness and brightness of newsprint. They found maximum yellowing by the near-ultraviolet light, and photobleaching by blue light. Prolonged exposure (approx. 14 hours) to visible light also resulted in yellowing of the newsprint, whereas samples preexposed before the tests to what the authors identified as "white light" did not reproducibly undergo significant changes when exposed to light above 400 nm (see chap. 3 for more details). Calculations by the authors showed that degradation of the newsprint was a linear function of the light intensity. The number of photons irradiating the paper samples cannot be determined from the data provided in this study; thus, a recommendation for photoflash exposure cannot be made. These results underscore the importance of knowing the history of the exposure of ligneous papers to light to be able to predict the effects of additional light exposures.

> **Conservators' Note:** *These results illustrate the fact that the responses of ligneous papers (including newsprint) to light will vary considerably, depending on the composition of the paper, the method of manufacture, the acidity of the paper, and its prior history of light exposure. In general, papers containing lignin will be noticeably less tolerant of exposure to photoflash light sources than pure cellulose papers are. The amount of near-ultraviolet light emitted by the source will be critical in determining the allowable lifetime exposure to flash. The maximum amount might be only about 10^3 flashes from a xenon studio strobe setup without color-correcting filters, or one*

to two hundred photocopies. It should also be noted that several hours of exposure to fluorescent lamps lacking ultraviolet-blocking sleeves or several minutes of exposure to a mercury arc used for ultraviolet photography could deliver a dose of near-ultraviolet light equivalent to 0.01 of the display lifetime exposure.

Research is under way to determine if using ultraviolet-screening compounds and reducing agents to treat handsheets of lignin-containing papers will protect them from discoloration and photochemical degradation. To test the effectiveness of the additives, Trichet and coworkers (1996) irradiated paper samples with ultraviolet light of wavelengths above 300 nm. They found that some additives confer partial protection from discoloration; however, the reducing agents appear to be quickly consumed during irradiation.

Conservators' Note: *These results suggest that ligneous papers containing the more effective ultraviolet-screening materials may be as much as an order of magnitude less sensitive to yellowing by exposure to ultraviolet light. Because application of the screening agents to papers was still experimental as of the mid-1990s, papers in museum and archival collections up to that time probably should not be presumed to be protected in this manner.*

Wood

The effects of light on wood are very complex because of the complex nature of wood itself, which consists of cellulose, hemicelluloses, lignin, colorants, and so on. Ultraviolet light does not penetrate, at least initially, more than about 75 μm into the wood, so the effects are confined to the surface region (Hon 1991). Different species of woods respond very differently to outdoor weathering. For example, redwood darkens significantly for the first several months, while Western red cedar is bleached. In general, the time frames used for outdoor experiments on which these and similar observations are based, and the lack of detailed information in published studies on the light exposures used, prevent the data from being useful in the current context.

A few laboratory investigations were performed in the mid-1990s on well-characterized woods. Choudhury and coworkers (1995) looked at the fluorescence induced by ultraviolet (365 nm) or blue (450–490 nm) light on very thin sections (30 μm) of fir wood to compare the photodegradation of lignin in these samples with that of ground wood pulp from the same source. They demonstrated an apparent photochromic effect that is similar in the two types of materials and is also seen in pulp reduced by sodium borohydride. When the thin sections of wood were chemically reduced, most of the blue-light-induced fluorescence was lost, suggesting that the absorbing chromophore was lost. In the absence of additional information on sample preparation and the light sources, the implications for possible effects of exposure of this wood to flashlamps are unclear.

In the course of investigating the protective effects of several antioxidants on the response of maritime pine to artificial sunlight from a Xenotest apparatus, Castellan and coworkers (1996) observed that unprotected wood samples darkened unacceptably after only about 2 hours of experimental exposure. The light source was not described in detail. If it is similar to an Atlas Fade-Ometer xenon arc, the observations suggest that this pinewood has a sensitivity to near-ultraviolet and visible light only slightly less marked than that of lignin-containing paper. Some of the antioxidants tested were ineffective, whereas others provided partial protection from discoloration under the experimental conditions used.

French researchers have reported some useful observations on the discoloration of lowland white fir and of European oak woods under various conditions (Dirckx et al. 1992; Mazet et al. 1993). They exposed well-characterized samples to either ultraviolet or visible light from a mercury-vapor lamp. In the amount of time that the ultraviolet light caused noticeable yellowing of the fir wood, visible light caused imperceptible (but experimentally detectable) darkening. Infrared spectroscopic data obtained on irradiated fir wood and on lignin were similar, suggesting that the changes in the wood were due to reactions of lignin. Significant changes in the wood were observed after 32 hours of irradiation (see chap. 3 for calculations).

> **Conservators' Note:** *These results suggest that about 2×10^5 flashes from a studio strobe source without color-correcting filters would be required to deliver 0.01 of the photons that the experimental samples received. This indicates that moderate flash photography of fir wood should not be problematic.*

The responses of European oak sapwood and heartwood were also investigated by these researchers. Oak heartwood (the older, plugged phloem part of the tree trunk) contains much larger deposits of extractable colored substances, such as gums and tannins. Ultraviolet light caused detectable changes in the appearance of these woods in less than one-tenth of the time in which they occurred in fir. The responses of the oak heartwood and sapwood samples were quantitatively different, and the initial difference in appearance continued to increase for the first 100 hours of exposure. Irradiation of heartwood with visible light caused bleaching at about 580 nm, as well as some of the changes in the infrared spectra that were observed for the oak sapwood and the fir. These changes occurred during very long exposures to a powerful visible light source.

> **Conservators' Note:** *The differential response to light should be kept in mind when display under natural lighting conditions is considered for objects containing both oak heartwood and sapwood; such conditions could result in a changed overall appearance after long display times. Moderate amounts of flash photography are not likely to*

> *significantly affect the appearance of objects containing oak woods. However, ultraviolet photography of artifacts that have exposed components of both sapwood and heartwood and that have not previously received long light exposures may not be advisable.*

Another complicating aspect of the situation for wooden objects in museum and archival collections is that most of these objects are stained or coated with colorants, resins, waxes, paints, or other coatings of various transparencies. In many cases, the effect of photographic or reprographic light on a wooden artifact might more reasonably be considered as the effect on the coating material rather than on the underlying wood. This problem does not appear to have been addressed directly in the literature. A cautious approach might be to assume that wooden artifacts have a sensitivity no greater than that of ligneous papers, as long as they have not been coated with materials of greater light sensitivity.

Gums and Natural Resins

Gums are polysaccharides isolated from plants and (partially) purified for use as binders and extenders for paints and inks. They occasionally have been used without additives to provide a shiny coating on pigments, the paper support, or both. At the beginning of the twentieth century, gum arabic (gum acacia) was also used with bichromate salts as a photographic print emulsion (Scopick 1978). These natural products consist of straight and branched polysaccharide chains containing a variety of monomeric sugars and sugar acids. The polysaccharides do not absorb significant amounts of visible or near-ultraviolet light and should not be adversely affected by photographic flashlamps.

Gums frequently contain small amounts of other plant materials, however, such as starch and proteins. Some of these other compounds are more likely than the polysaccharides themselves to be responsible for discoloration of the gums if they undergo photodegradation. Gamboge is a classic example of a colored mixture of gum and resin. The yellow color of gamboge is due to gambogic acid, a complex heterocyclic isoprenoid derivative with acid, ether, hydroxyl, and carbonyl substituents (Mills and White, 1994a). Reports of research specifically addressing the lightfastness of the colored components of gamboge were not located.

Natural resins are typically alkyd polyesters consisting of alcohols with two or three hydroxyl (–OH) groups esterified to dicarboxylic acids. They may also contain a small percentage of long-chain organic acid substituents (Seymour and Carraher 1992, chap. 6). Natural resins can be highly branched because additional cross-linking occurs as the oil component (the long-chain fatty acid) dries and the resin hardens. They contain large numbers of carbonyl groups and double bonds, so they absorb in the near ultraviolet and can yellow over time. A wide variety of natural resins—shellac and dammar resin are two examples—has been used in art objects (Mills and White 1994b).

De la Rie (1988a, 1988b, 1993) has extensively studied the photodegradation and stabilization of resins used as picture varnishes. Long exposures to the xenon arc in a fading apparatus caused significant yellowing of dammar resin films; about 80% of the yellowing was caused by near-ultraviolet wavelengths. The changes appear to occur mostly in a thin surface layer of the varnish that absorbs the damaging ultraviolet light and protects the underlying varnish and paint somewhat. About 10^6 flashes from a xenon strobe lamp would cause less than 0.01 of the change observed in these experiments (see chap. 3). De la Rie also observed that if the resin films were heat-aged at 80 °C for a few hundred hours before being exposed to the xenon arc, or if they were wet, major discoloration occurred after 20 hours of light exposure.

> **Conservators' Note:** *These data indicate that varnished paintings that may have been stored in very hot, moist places for long periods should probably not be subjected to extensive flash photography. Because paintings in this condition will most probably have had the old, discolored varnish removed and replaced, however, this may not be a concern for museum personnel.*

Slight bleaching of yellowed resin by light of wavelengths greater than 400 nm was also observed but not quantitated (de la Rie 1988a). It occurred to a lesser extent than the yellowing and can account for areas of darker yellow varnish on those parts of an object protected from long-term exposure to light that contained near-ultraviolet wavelengths. This phenomenon does not alter the conclusions reached regarding the relative safety of moderate flash photography.

De la Rie (1993) also studied the removability of dammar resin and some synthetic resins as a function of the polarity of the solvent used and the length of time that the resin films had been exposed to a xenon arc with filters to simulate sunlight through window glass. The resin required polar solvents for solubility after about 50 hours of light exposure and was not protected by inclusion of hindered amine light stabilizers unless ultraviolet light was excluded.

> **Conservators' Note:** *Because photographic flash sources do not radiate the lower-wavelength (higher-energy) ultraviolet component of sunlight, these experimental results should not raise additional concerns for the safety of natural resins exposed to photoflash.*

Synthetic Polymers

Most synthetic polymers with which conservators deal—for example, synthetic textile fibers, plastics, resins, varnishes and other coatings, and adhesives—are not, or should not be, colored materials in their pure state (McKellar and Allen 1979, chap. 1; Rånby and Rabek 1992); that is, they should not absorb visible light, based on the chemical formulas for the pure compounds. In fact, they also should not absorb near-ultraviolet wavelengths above about 350 nm (exceptions are mentioned in the following discussions). When polymers do absorb near-ultraviolet or visible light, it is often due to the presence of chromophores. These are molecules or substituents that impart color to a material. They may become incorporated inadvertently into the polymer during or after manufacture, or they may be added to the formulation on purpose. The chromophores in a polymer may be present as integral units of the backbone chain, as end groups, and/or as side groups connected to the backbone atoms. They may be regularly or irregularly spaced along the polymer molecules. They may or may not be involved in any cross-linking of the chains, or they may be present as separate molecules. Chromophores generally contain conjugated double bonds.

Some types of polymers are synthesized or modified during manufacture by a photoinitiated or photodriven process (Rånby 1995). These materials are by nature expected to stand up to moderate exposure to ultraviolet light. In most cases, the radiant energy and the wavelengths required to cause chemical alteration of a synthetic polymer far exceed the light emitted by flashlamps and reprographic light sources, so flash photography or photocopying of these materials should not be of concern. However, the presence of additives, impurities, or degradation products may change the sensitivity of a polymer to near-ultraviolet or visible light. Specific examples are mentioned in the following paragraphs (the separate issue of polymer degradation sensitized by dyes included in the polymer is discussed under "Polymeric Materials Containing Colorants," pages 78–102).

Photoinduced modifications and breakdown of polymers can proceed by one or more of several processes. A distinction may be made between photodegradation (in the absence of oxygen) and photooxidation (in the presence of oxygen) of these materials. This is discussed in Grassie and Scott (1985, chaps. 3, 4) and in Allen and Edge (1992d). Elementary aspects of the chemistry involved have been described by Wayne (1988, chap. 6). More-detailed presentations can be found in McKellar and Allen (1979) and in Rabek (1987b). Allen and Edge (1992d) and Rånby and Rabek (1992) have published reviews of this information.

At least three different processes leading to photoinduced modifications or breakdown of polymers, all involving radical formation, may occur under appropriate conditions.

1. Absorption of light can directly cause cleavage of carbon–carbon or carbon–hydrogen bonds in the polymer chain. Two radicals are formed, one or both of which subsequently react to cause cross-linking or further scissions.
2. When carbonyls (>C=O groups) are present in the chain, a break in the chain can occur at the carbonyl carbon; again, two radicals are formed, causing subsequent reactions, and CO may be evolved. In the presence of oxygen, a highly active peroxide radical is formed. It can take a hydrogen atom from elsewhere in the polymer (hydro-

gen abstraction) to form a peroxide group on the polymer and leave another reactive radical where the proton substituent (–H) had been. There are now three reactive groups in the polymer (the original radical next to the break at the carbonyl group, the peroxide, and the new radical somewhere else), and all of them can undergo additional reactions.

3. When a carbon with at least one hydrogen atom attached to it is located three atoms along the chain from a carbonyl group, a radical reaction can occur involving the carbonyl carbon and this hydrogen. The reaction results in chain scission and the formation of a double bond between the two carbons on one side of the break. The carbonyl is regenerated at the other side of the break in the chain. No radical is left after this reaction. Much less commonly, the photoexcited carbonyl group will abstract a hydrogen directly from elsewhere in the polymer to form the reactive peroxide group >C–O–OH and another radical along the polymer where the proton substituent was removed.

The second of these reactions appears to be the most common. For example, it can make a significant contribution to the degradation of polypropylene when the carbonyl group is present as an impurity in very small amounts, such as 1 in 200 carbons (Faucitano et al. 1996). In most cases, the wavelengths of light that are effective in causing photodegradation and photooxidation are well below 360 nm, and the irradiances required are normally high. Thus, the exposure of most synthetic polymeric materials to photographic flashlamps, or to photocopier light sources, is not likely to cause significant deleterious changes. Exceptions may occur, as the following examples show.

Polymer materials that have received extensive exposure to sunlight may already be highly oxidized and somewhat degraded. Some of these materials may, as a consequence, contain chromophores that absorb in the visible region, so the polymer is then sensitive to longer-wavelength radiation. Polymeric objects that are intended to be photodegradable contain photosensitizers, which initiate and catalyze the degradation processes (e.g., environmentally friendly, disposable polystyrene cups). They may have been purposely designed to self-destruct completely in 20 hours of sunlight (Allen and Edge 1992a). Exposure of these objects to ultraviolet light should be avoided, although visible light may not be harmful.

Some authors have summarized photochemical data on several particular polymers or types of polymers. McNeill (1992) has presented a very readable introduction to polymer degradation, including a section on photodegradation. Rånby (1993) classified certain groups of polymers according to their photostabilities. Poly(tetrafluoroethylene) (PTFE) and poly(methyl methacrylate) are considered highly stable, without any photostabilizers added. Moderately stable polymers, which should last a few years outdoors without stabilizers, are poly(ethylene terephthalates), polycarbonates, poly(vinyl fluoride) and poly(vinylidene fluoride). The poorly photostable group of synthetic polymers in Rånby's scheme, which require stabilizers for outdoor use, are polyolefins (that is, hydrocarbon polymers), polyurethanes, polystyrene, poly(vinyl chloride), polyamides, diene rubbers, and what the author calls synthetic polymeric coatings. In his discussion of the photoinduced degradation of some of the members of this last group, however, Rånby indicates that ultraviolet, but not visible, light caused the changes observed.

Allen and Edge (1992d) have published the near-ultraviolet absorption spectra of several types of synthetic polymers. Of those presented, only poly(ethylene-2,6-naphthalate), which is a less common polyester; a polyurethane made from 4,4′-diphenylmethane diisocyanate; and poly(ether sulphone) have significant absorption above 350 nm. In an adjacent table, however, the authors indicate that 364 nm light will also activate changes in poly(vinyl chloride) and poly(vinyl acetate). Faucitano and coworkers (1996), in a discussion of hindered amine light stabilizers for polymers, state that aliphatic polyamides, poly(ethylene terephthalate), and polycarbonates without stabilizers underwent photo-oxidations when irradiated with light at about 360 nm (probably from a mercury arc). The data they provided either did not indicate what wavelengths were active or showed results for polymers irradiated with shorter wavelengths than are relevant for this discussion. Rånby and Rabek (1992) presented photochemical degradation mechanisms in detail and also listed quantum yields for chain scission of certain polymers irradiated at specific wavelengths. Of the compounds listed in their table 3, only poly(ethylene terephthalate) and cellulose acetate are identified as sensitive to wavelengths greater than 340 nm.

Andrady (1993) reviewed the possible effects of ultraviolet-A radiation on some polymeric materials. In a tabulation of discoloration effects, he indicated that only acrylonitrile-butadiene-styrene copolymer was sensitive to discoloration by exposure to wavelengths greater than 355 nm. In one of the figures in the publication, however, he shows that poly(vinyl chloride) also had a slight tendency to yellow when exposed to light between 375 and 405 nm, although the major effect was caused by wavelengths less than 360 nm. In providing still another list of wavelengths that were effective in causing polymer degradation, Seymour and Carraher (1992, 398) indicate that polypropylene can be degraded by 370 nm light, whereas the other materials listed are not sensitive to ultraviolet light with an energy this low.

Fornes's research group has shown that high concentrations of atmospheric pollutants can increase the deleterious effects of ultraviolet light on synthetic polymers. Fornes and Gilbert (1991) summarize the effects of the acidic smog components nitrogen dioxide, NO_2, and sulfur dioxide, SO_2, on various polymeric materials, including latex and alkyd films. They exposed the materials to very long periods of irradiation in xenon-arc instruments. These exposures would be the equivalent of billions of flashes from photoflash lamps. Their observations suggest that exposure to flash does not pose a significant threat to the materials they investigated. Their results also show, however, that the presence of calcium carbonate, $CaCO_3$, in the films can cause significant morphological and color changes that are not observed in the absence of the extender.

This brief summary illustrates the fact that individual research reports and reviews have based their conclusions on studies of selected groups of polymers available at the time of the investigation. The lists of compounds sensitive to light are never exhaustive. The absence of a polymer from such a list may not necessarily indicate its general stability but could be due to the fact that it or members of its class were not included in the study or were not available at the time of the research. The samples tested may have been manufactured with different procedures or ingredients from those used in the production of a particular museum object. The technology of polymer manufacture is constantly evolving. A particular compound that was light sensitive a few years or decades ago may not have the same sensitivity in its current formulation, and vice versa. Specific examples based on numerical data are discussed in the following sections.

Polyolefins

The results published by several different investigators all indicate that high- and low-density polyethylenes (HDPE and LDPE), polypropylene, and related polymers are stable to near-ultraviolet and visible light. This is true whether or not they contain additives, such as metal ions or antioxidants, for example. These polymers can be expected to be essentially unaffected by exposure to photoflash or photocopier light sources (see corresponding section of chap. 3 for details on such reports).

An example of a modern polyolefinic object with which conservators might be confronted is Tupperware, commercial containers made from polyethylene. They are beginning to appear in museum collections and in composite objects, as well as being used as storage containers in the museum environment. Van Oosten and Aten (1996) have compared naturally aged Tupperware to samples of new Tupperware products and of HDPE and LDPE artificially aged by exposure to ultraviolet light from a filtered xenon-arc source for hundreds of hours in a Suntester chamber. They observed changes in the oxidation state of the polymers after 100 hours of exposure. These changes appeared to continue upon further irradiation.

> **Conservators' Note:** *The light levels used by van Oosten and Aten were several orders of magnitude greater than those emitted by photographic light sources. The data do not provide evidence to suggest that flash photography needs to be restricted for these materials.*

Polyesters

Two main types of polyesters are considered here. The first are the polycarbonates, which are polyesters of carbonic acid and a dihydroxyphenyl compound, most commonly bisphenol A (Seymour and Carraher 1992, sec. 7.5). These polymers are considered to be very stable to ultraviolet light. For example, Lexan, which has been used as an ultraviolet-blocking glazing in museums, is a bisphenol A polycarbonate (Andrady, Searle, and Crewdson 1992). Studies of the photodeterioration of these polymers are typically conducted with light sources that include wavelengths below 320 nm (Shah, Rufus, and Hoyle 1994) or with the plastic exposed for years to unfiltered sunlight. Polycarbonate products intended for use where they will receive ultraviolet light exposure are usually stabilized with an ultraviolet absorber such as a benzotriazole; this type of additive is what gives Lexan its slight yellowish tint. Andrady, Searle, and Crewdson (1992) have obtained activation spectra for yellowing induced in stabilized and unstabilized bisphenol A polycarbonates. Wavelengths above 360 nm do not cause changes in the stabilized polymer. The contribution of ultraviolet wavelengths above 355 nm to the changes observed in unstabilized polycarbonate were trivial compared to the effects of ultraviolet light below 320 nm.

> **Conservators' Note:** *It can be concluded that polycarbonate will not be endangered by exposure to photoflash or reprographic light sources.*

A second group of polyesters of interest to conservators are copolymers of terephthalate and either saturated or unsaturated short-chain aliphatic compounds, such as poly(ethylene terephthalate), also known as PET. Dacron and Mylar are well-known fiber and sheet forms, respectively, of this polyester (Seymour and Carraher 1992, chap. 7). Some authors' evaluations of published results have led them to suggest that these materials will not be significantly affected by light of wavelengths greater than 320 nm (Imagi et al. 1987; Rabek 1987b). Other researchers have concluded that near-ultraviolet light may also contribute to photodegradation. An analysis of the data reported by McKellar and Allen (1979, 147–53) indicates that the experimental doses used exceeded the equivalent of 10^8 flashes from a xenon studio strobe (see chap. 3 for more details).

> **Conservators' Note:** *It can be concluded that PET polyester that has not been degraded should not be adversely affected by flash photography or reprography. However, poly(ethylene terephthalate) that has already been exposed to hundreds or thousands of hours of ultraviolet light does have altered spectroscopic properties (McKellar and Allen 1979, 147–53) and may be somewhat more susceptible to further change if it receives additional near-ultraviolet radiation. The presence of colored impurities or dyes may alter the response of the deteriorated polymer to near-ultraviolet or visible light.*

Some less common polyesters absorb near-ultraviolet or blue light and are, therefore, much more sensitive to light of these wavelengths than polycarbonates or PET (Allen and Edge 1992d). Examples are linear aliphatic unsaturated polymers and poly(ethylene 2,6-naphthalate), also known as PEN. Relatively short exposures of this polymer to near-ultraviolet or blue light sources cause major changes. Approximations based on the available experimental details suggest that about 10^3 flashes from a xenon studio strobe without ultraviolet filtering could be the maximum allowable for PEN (see chap. 3). Start-up of major industrial production of PEN in the mid-1990s suggests that the vulnerability of this polymer to near-ultraviolet light may become better known (*Chemical and Engineering News* 1994).

> **Conservators' Note:** *In general, the chemical nature of the polymers in unusual, colored, or degraded polyesters should be ascertained before the light exposures to which they will be subjected are chosen. Also, records of their exposure to light should probably be maintained.*

Polystyrenes

In polystyrenes, the phenyl group is a side group in the polymer and not part of the backbone chain. The pure polymer would not absorb at wavelengths greater than 300 nm, and most research on photodegradation or photooxidation has been performed with light sources containing ultraviolet light of shorter wavelengths than this.

It is obvious from the published data that the relatively short near-ultraviolet light exposures during flash photography or photocopying are not likely to be deleterious to polystyrene. Additionally, polystyrene is used extensively in mixed polymer formulations. Studies of different mixtures containing this polymer all indicate that polystyrene is very stable to visible and near-ultraviolet light in a variety of formulations (see chap. 3 for summaries of research on various materials).

Some polystyrene products have been purposely designed to be degraded by exposure to sunlight, for example, disposable food containers. Objects made of these materials may occasionally be found in museum collections. Because the sensitizers incorporated into the polymer usually absorb ultraviolet light between 300 and 400 nm (McKellar and Allen 1979, chap. 3), and because they are designed to degrade on a timescale of months or years outdoors, a modest number of flash exposures of a fraction of a second should not be expected to have a deleterious effect.

> **Conservators' Note:** *If polystyrene objects of this type already appear to be partially degraded, the presence of a degradable formulation should be suspected. In this case, especially if the objects have a known history of exposure to sunlight or ultraviolet radiation, repeated exposures to photoflash light sources without complete ultraviolet blocking might best be avoided.*

Polyamides

Nylons are the most common synthetic polyamides. (Proteins are also common polyamides; the fibrous proteins wool and silk, which are of particular interest to conservators, are considered earlier, in the section on natural fibers.) Although ultraviolet light below 300 nm is the most effective in causing photodegradation of undyed nylon-6,6 (also referred to as nylon 66) yarn, near-ultraviolet and blue light can also induce damage. For example, Katsuda and coworkers (1996), in a study of undyed and dyed fabrics, report that although the appearance of undyed cotton and polyester was not affected by exposure to light of wavelengths above 370 nm, nylon was maximally affected at this wavelength (see chap. 3 for details).

> **Conservators' Note:** *Despite the possibility of near-ultraviolet and blue light causing photodegradation of nylons, the evaluation of experimental data obtained following different procedures suggests that undyed nylon could tolerate on the order of 10^5 xenon strobe flashes. Moderate exposure to photoflash sources is not likely to cause detectable change to this polymer. However, irradiated nylon may have a decreased capacity to take up certain dyes (Stowe et al. 1973).*

Information on other polyamides (see chap. 3) suggests that they would also be only moderately sensitive to photoflash or reprographic light exposure, able to tolerate tens of thousands of flashes during their display lifetimes. An exception to this behavior is a polyamide that contains a thiazole; it may be affected by about 10^3 flashes (Johnson, Tincher, and Bach 1969). The light sensitivity of any particular polyamide will depend on the composition of the polymer and not on its general features.

Dimers of some aromatic ketones, such as coumarin or umbelliferone, can be copolymerized with, for example, amides to form derivatized polyamides. In this state, they can be photochemically cleaved by ultraviolet light into a variety of products, such as stilbene derivatives, that remain part of the polymer (Hasegawa and Saigo 1990). These types of derivative polymers are becoming more common as engineers and materials scientists attempt to incorporate properties such as fluorescent brightening directly into the polymeric materials. Conservators should probably expect to find unexpected combinations of properties in contemporary plastic objects, and this might in some cases affect their responses to light.

Polyacrylates and polyacrylics

Both poly(methyl acrylate) and poly(methyl methacrylate), commonly known as PMA and PMMA, respectively, are considered in general to be very stable to light above 300 nm. Most studies of their photostability have been performed using shorter-wavelength ultraviolet radiation (McKellar and Allen 1979, 85–93; Rånby 1993). Feller, Curran, and Bailie (1981) have summarized research on the effects of accelerated photochemical aging of poly(alkyl methacrylate) films. Their main concern has been to determine if exposure of these varnish components to display lighting will result in cross-linking and insolubility of the varnish layer. The data suggest that they would be very stable to flash exposures (see corresponding section of chap. 3). The addition of light stabilizers, such as Tinuvin, to the polyacrylates, currently a common practice, would further reduce the probability of damage from exposure to flash.

In analogy with the covalent incorporation of fluorescent whitening agents into polyamides, the feasibility of covalently binding ultraviolet stabilizers to methacrylate polymers has been investigated by Nelson and Wicks (1982). The purpose of this study was to bind the stabilizers so they could not migrate out of the acrylic polymer when it was used in conservation applications. The stabilizers were found to be as effective when they were covalently bound as when they were included in the polymer formulation as separate molecular entities.

Another important group of polymers based on acrylic acid derivatives is exemplified by the polyacrylonitriles. These include acrylic polymers used as latex binders in paints and coatings, acrylic sheeting such as Plexiglas, and acrylic fibers made of polyacrylonitrile, such as Orlon. A study of a latex coating for paper that included a small amount of acrylic polymer is discussed under "Pigment-coated papers," pages 101–2. The acrylic component in that study (Mailly et al. 1997), however, was not specifically considered as a photosensitive ingredient. Whitmore, Colaluca, and Farrell (1996) investigated the cause of turbidity development in artists' acrylic paint. They determined that crystallization of poly(ethylene glycol) was responsible for the effect and that this polymer remained in the paint because of its high affinity for the acrylic matrix. This research is continuing; no results of photochemical studies have yet been published. Typically, the photochemistry of polyacrylonitrile has been studied at wavelengths below 300 nm (McKellar and Allen 1979, 93–95). The undyed or unmodified polymer can probably be assumed not to be adversely affected by exposure to photoflash.

> **Conservators' Note:** *In general, polyacrylics and closely related polymers, in the forms normally encountered in collections (Plexiglas, varnishes containing ultraviolet stabilizers, Orlon fibers), may be assumed to be relatively stable to photoflash exposure.*

Poly(vinyl —) polymers

Pure poly(vinyl acetate) and poly(vinyl alcohol) have negligible absorption above 300 nm. Ultraviolet light below 300 nm causes changes in their absorption spectra. They will develop absorption bands above 350 nm after they have received many hours of radiation at the shorter ultraviolet wavelengths. Flash photography with filtered xenon light sources and photocopying would not appear to pose a threat to either of these materials.

Poly(vinyl chloride), commonly known as PVC, and other similar polymers also should not absorb light above 300 nm. In most formulations, however, manufacturing processes result in the presence of impurities that will absorb in the near ultraviolet (e.g., Rabek 1987a; Allen and Edge 1992d). These inclusions may be responsible for rapid discoloration when PVC is exposed to light of wavelengths greater than 365 nm (Rånby and Rabek 1992). The situation for this polymer is made more complicated by the fact that it typically contains large amounts of additives (up to 40% by volume), such as plasticizers, fillers, colorants, and so on.

Specific studies of the light sensitivity of PVC have suggested that it would be only moderately sensitive to photoflash, even when not stabilized, provided wavelengths below about 385 nm were removed from the source. For example, calculations based on the data of Martin and Tilley (1971) indicate that at least 10^5 flashes could be tolerated (see chap. 3 for details).

More recently, Andrady and coworkers determined action and activation spectra for photochemical changes and discoloration of some PVC formulations. The results, summarized by Andrady (1993, 1997), suggest that the addition of rutile titanium dioxide to poly(vinyl chloride) makes it significantly more sensitive to discoloration by ultraviolet light of wavelengths less than or equal to 340 nm, but not greater. Unpigmented PVC was found to be very slightly sensitive to yellowing when irradiated by light with wavelengths of 370–410 nm as well. Use of empirical equations derived by these researchers for change in yellowness index indicates that about 10^5 flashes could be tolerated by this polymer (see chap. 3 for details).

> **Conservators' Note:** *Although poly(vinyl chloride) objects are generally considered to be sensitive to light, this class of polymers should be able to tolerate moderate exposures to photoflash lamps, only if ultraviolet wavelengths have been blocked. Other poly(vinyl —) polymers should not be unduly sensitive to photoflash.*

Polyurethanes and related polymers

Polyurethanes are polymers containing amide groups and also ester, ether, or thioether linkages in the polymer chain (McKellar and Allen 1979, chap. 1). They may have benzene rings in the polymer backbone (aromatic polyurethanes) or not have ring structures (aliphatic polyurethanes); there is usually extensive hydrogen bonding. Because of the large variety of structural possibilities, the polyurethanes have a wide range of properties (Seymour and Carraher 1992). Photochemical studies of polyureas and polyurethanes have usually been conducted with ultraviolet wavelengths below 340 nm (e.g., Hoyle, Shah, and Moussa 1996), despite the fact that a few of these compounds have minor absorption bands or shoulders at about 350 nm. Prolonged, high-intensity irradiation of polyurethanes with the higher-energy ultraviolet light can cause extensive yellowing (Rabek 1987b; Allen and Edge 1992d). Effects of near-ultraviolet and visible light on the yellowed materials do not appear to have been investigated quantitatively.

> **Conservators' Note:** *Data on the effect of near-ultraviolet and visible light on polyurethanes is lacking. However, there is no indication in the literature that polyurethane polymers not already yellowed would be very sensitive to photoflash light sources.*

Other polymers

Silicone polymers, with the exception of silicon hydrides, have been shown to be very stable to near-ultraviolet light (Lacoste, Israëli, and Lemaire

1996) and are recommended for use in severe environmental conditions. Flash photography of these materials should not be of any concern.

Epoxy resins should not absorb light of wavelengths above 300 nm when they are properly formulated. If the ratio of hardener to resin is not within the guidelines for the particular formulation, however, or if the amount of plasticizer or accelerator is not appropriate, enhanced yellowing by near-ultraviolet light may be observed (Tennent 1979). The long times required for yellowing to occur (e.g., several hundred hours of exposure to a Microscal lamp) suggest that radiation from photoflash sources would probably not cause a significant change in appearance. A later study of more than two dozen different epoxies suggests that formulations that do not yellow in the dark will also be very stable to light (Down 1986). Because the equivalent of years of light exposure is required to discolor these adhesives significantly, they should not be affected by photoflash exposures.

Polysulfones, containing $-SO_2$ groups and benzene rings in the polymer backbone, absorb some near-ultraviolet light in the 320–400 nm range. They have been shown to yellow if exposed to ultraviolet light (Allen and Edge 1992d). A few hours of exposure to xenon-arc light in a weathering apparatus significantly decreases the extensibility of sheets of this polymer. The exposure times used in this study were equivalent to more than 10^4 flashes from a xenon flashlamp. As with polyurethanes, however, quantitative investigations of the response of yellowed polysulfones to near-ultraviolet- or visible-light irradiation do not appear to have been made.

The perfluoronated sulfonated polymer Nafion is used as a support for photochemical studies employing xenon sources (Niu et al. 1988) and would not be expected to be altered by flashlamps.

Conservators have been interested in parylenes because of their possible role as protective coatings. Bilz and Grattan (1996a, 1996b) have shown that parylene N (polyxylene) and parylene C (polymonochloro-*p*-xylene) can be photooxidized by a fluorescent light source filtered to block ultraviolet light of wavelengths less than 380 nm. About 10^3 and 10^4 hours of exposure, respectively, to this source are required to obtain detectable amounts of carbonyl groups in the two parylenes. These exposures are equivalent to about 4×10^7 flashes or 4×10^8 flashes, respectively, which suggests that flash photography or photocopying of objects coated with these polymers would probably not harm the coatings.

Cellulose ethers have been used in conservation as adhesives and consolidants for the better part of this century, due to their ease of use and their reversibility. Cellulose ethers and esters were also popular materials for films and coatings until the past few decades (Horie 1989). Unfortunately, they tend to be unstable in the dark as well as in the light. Several researchers have investigated ultraviolet-light-indexed deterioration due to of materials such as cellulose nitrate, but there appears to be a dearth of information on the effects of light for wavelengths greater than 365 nm (Selwitz 1988, 22–25, 38–39).

Polymeric Materials Containing Colorants

Many tests of the lightfastness of colorants in polymeric materials have been conducted by the dye and pigment industry to evaluate synthetic colorants for use in textiles and coatings. The research is driven not only by the desire to avoid marketing products whose colors are fugitive but also by the phenomenon of phototendering, that is, the photosensitization of fiber degradation by the colorant (Allen and McKellar 1980a; McLaren 1983). The results of these investigations have shown that many physical and chemical factors can influence the effect of light on dyed polymers. It has been known for some time that changes in relative humidity and the moisture content of dyed textiles can influence lightfastness more than changes in temperature (e.g., Trotman 1970). Herbst and Hunger (1993) have discussed several of these factors from the point of view of industrial pigment production, testing, and use.

Unfortunately from the perspective of this review, most experiments on the lightfastness of colorants in polymeric materials have been performed by exposing test samples to polychromatic light sources such as daylight or unfiltered lamps, which emit a significant portion of energy in the ultraviolet-A region below 365 nm. Also, exposures for the shortest times employed were often sufficient to fade Blue Wool no. 3 or higher. These conditions did not often provide the numerical results needed to calculate the effects of near-ultraviolet- and visible-light flash exposures on the most sensitive colored materials.

The distinctions among fundamental photochemical research, experiments on light stability performed with polychromatic light sources, and comparative studies of lightfastness of colorants have been stated very clearly by McAlpine and Sinclair (1977). They stress the fact that, with the exception of fundamental photochemical investigations, the results are qualitative even if the experimental protocol employs standardized equipment and procedures. In some cases, publications based on comparative studies have provided enough information to allow *approximations* to be made of the possible effects of flash or photocopier light exposures on art and archival materials containing colorants.

Selected examples of some useful findings are discussed here. Many of the evaluations were made using guesstimates for numerical data that were not provided in the publications. Calculations performed with these numbers give semiquantitative results at best and usually provide only vague or order-of-magnitude values. Conclusions based on these calculations should be regarded as approximate guidelines only. For obvious reasons, this review of the vast literature on dyed polymeric materials could not be exhaustive, and the data chosen for presentation do not necessarily imply that the particular colorants discussed are commonly found in museum collections or that they are the most sensitive ones that conservators may need to consider.

Additional uncertainty is caused by the fact that the *rate* of fading of colorants may change with the extent of exposure. Giles and his colleagues published extensively on this aspect of lightfastness (e.g., Giles, Johari, and Shah 1968). They identify five different patterns of fading kinetics. In three of them, the rate of fading decreases as total exposure increases. In only one case does the rate of degradation of the colorant increase with time of exposure, and few examples of this behavior have been reported.

Textiles containing natural dyes

Authors of books and reviews on dyeing textiles with natural dyes usually provide qualitative comments on the lightfastness of these materials (e.g., Cannon, Cannon, and Dalby-Quenet 1994; Liles 1990; Grierson, Duff, and Sinclair 1985b). In earlier studies on the sensitivity of dyed fabrics to light, data were frequently obtained using daylight or incompletely defined light sources and were collected over long exposure times. Results of experiments were often described qualitatively or reported as rankings with respect to Blue Wool standards or as changes on a gray scale.

Comparative studies of Western dyes on wool and cotton

Padfield and Landi (1966) summarized early studies of the lightfastness of natural dyes. They present an excellent perspective for a balanced approach to the problem of dye fading in displayed textiles and to the design and interpretation of more recent experimental work. They also investigated the effects of light from fluorescent lamps (with or without filters to remove ultraviolet or blue and green light) on many natural dyes on wool or cotton cloth with a variety of mordants (see chap. 3 for details). The most sensitive dye-fabric combinations they investigated were brazilwood and turmeric on wool mordanted with aluminum, and logwood on cotton, all of which were less stable than Blue Wool no. 1. Comparison of their results with and without ultraviolet wavelengths blocked shows that several natural red and yellow colorants on wool and cotton may be particularly sensitive to photodegradation by near-ultraviolet light.

In his review of the fading of dyes on textiles, Gallotti (1994) presents a straightforward analysis of some of the simpler photochemistry and a good overview of other factors, such as mordants, fibers, and textile treatments that might possibly affect lightfastness. He mentions specifically the very fugitive nature of turmeric alone or mordanted with aluminum salts, but he does not provide data that can be used for quantitative assessments (see the corresponding section of chap. 3 for details on the dyed textiles).

Bowman and Reagan (1983) studied the lightfastness of turmeric, madder, or indigo dyes on cotton when ultraviolet and infrared wavelengths were removed from the radiation of incandescent, fluorescent, or quartz lamps used for display lighting. Their results suggest that turmeric is highly fugitive to visible as well as near-ultraviolet light on this textile. The madder-dyed cotton (mordanted with alum) appeared to be the most stable of the three dyes investigated under these lighting conditions.

Crews (1987, 1988, 1989) performed several comparative studies of the lightfastness of natural dyes on wool. The initial rate of fading of some red and yellow dyes on this textile was much greater than the rate of fading after significant color change had already occurred. In some cases, particularly for fustic and cochineal, the initial rate of fading also depended strongly on the mordant. Of the dyes investigated (cochineal, fustic, indigo, madder, turmeric, and weld), turmeric was by far the most fugitive. When mordanted with either alum or tin, turmeric on wool showed a perceptible fade after exposure to less than one AATCC fading unit of xenon-arc light in a weathering device. This observation suggests that about 100 flashes from a xenon studio strobe setup might be the maximum allowable over the display lifetime of wool dyed with turmeric (see chap. 3 for details). Old fustic and weld might be about one-fourth as light sensitive, depending on the mordant used. Madder was less fugitive than this, and indigo on wool without a mordant was relatively stable compared to the other dyes.

While investigating the effectiveness of several ultraviolet-blocking filters, Crews (1988, 1989) obtained additional data on the same dyes on wool. Exposures long enough to have caused significant fading without ultraviolet filtering were used to distinguish differences in the effectiveness of the filters. Turmeric-dyed wool was again found to be much more sensitive than the other samples, and none of the filters provided adequate protection from 80 AATCC fading units of exposure to xenon light. Results for weld were skewed by the fact that the ultraviolet present in the unfiltered light had yellowed the wool substrate, so bleaching of the dye was masked under these conditions. Fading was retarded to different extents by the different filters. An amber-colored filter that blocked a considerable portion of blue light between 400 and 550 nm, as well as near-ultraviolet wavelengths, was the most protective. An approximate calculation of allowable flash exposure over the display lifetime suggests that about 3×10^3 flashes of a xenon studio strobe setup, with filtering equivalent to the amber filter, could be the maximum for weld and madder on wool, depending on the mordant (see chap. 3). Filters normally used with this source probably transmit significantly more blue light than the amber filter used in this study, so exposure to xenon photoflash should probably be restricted to about 10^3 flashes total over the lifetime of the dyed textile. Turmeric may be many times more sensitive. The use of several approximations and extrapolation of the effects of a single, very long exposure to multiple, extremely short exposures make this conclusion tentative.

Duff, Sinclair, and Stirling (1977) studied the lightfastness of several traditional natural dyes on wool by irradiating samples of the dyed fabric in a Microscal apparatus. The most sensitive dyes they tested were sulfonated indigo and the yellow dyes from old fustic and Persian berries, which contain glycosylated flavonoids. The color changes they observed were reported in terms of b* (the CIEL*a*b* measure[1] of blueness or yellowness) rather than the overall color change, ΔE. Evaluation of the data (see chap. 3) suggests a very cautious estimate for total allowable numbers of flashes for these yellow colorants on wool of only about 100 flashes from a xenon flashlamp with ultraviolet filtering. However, this extension to filtered xenon sources is somewhat tenuous because the relative amount of near-ultraviolet light emitted at approximately 365 nm by the Microscal lamp is greater than that from a xenon flashlamp.

Grierson, Duff, and Sinclair (1985b) studied many dyes traditionally used in the Scottish Highlands to make woolen tartan cloth. The lightfastness of the freshly prepared samples was generally poor, being significantly less than that of indigo and logwood, for example. The red and purple lichen dyes were the most fugitive, quickly bleaching to colorless compounds in north daylight. A flavonoid yellow dye from heather twig tips faded fairly rapidly, according to the researchers, and a dark reddish purple from elderberry shifted to a green-black hue. No numerical data were provided in this publication. In another article, these authors (Grierson, Duff, and Sinclair 1985a) report that color changes equivalent to a ΔE greater than 15 were caused by exposing wool dyed with lichen purple to 20 hours in a Microscal apparatus. The information in this report suggests that again photoflash exposure should be limited to approximately 100 flashes over the lifetime of these very fugitive dyes on wool.

Conservators' Note: *The qualitative information in the studies reviewed here is rather well known. For example, turmeric, old fustic, orchil, and brazilwood (depending on the mordant) are likely to be extremely light fugitive. Evaluation of the data provides the numerical evidence that several red, purple, and yellow natural dyes on wool and cotton, if displayed at all, should not be photographed in the gallery, especially if they are in pristine condition. Studio flash exposures should be strictly limited to what is absolutely necessary. If they have been displayed, historic objects dyed with these compounds are likely to have already faded.*

Nineteenth-century American yellow dyes on wool

Crews (1982) investigated the lightfastness of a wide variety of natural yellow dyes that were used to dye wool in the United States in the nineteenth century. She showed that the mordant strongly influences the responses of these materials to artificial sunlight in a fading device. In general, alum and tin mordants caused much greater sensitivity to light than chrome, copper, or iron. The most sensitive dyes tested were turmeric, fustic, and marigold. The data reported in this study cannot be used to calculate photoflash exposure guidelines.

Red anthraquinone dyes on silk and cotton

Saito and coworkers (1988) compared the lightfastness of six naturally occurring anthraquinone dyes in solution or dyed on cotton and silk. The textile samples were weathered outdoors for 66 hours or exposed in a fading device for 100 hours, and the solutions were irradiated with a mercury-arc lamp, so that all samples received significant ultraviolet exposure. Purpurin (1,2,4-trihydroxyanthraquinone) was the most sensitive of the six colorants in solution and on both fabrics when exposed in a fading device. Quinizarin and quinalizarin on cotton and chrysazin (1,8-dihydroxyanthraquinone) on silk were the most fugitive in response to outdoor weathering. In all these cases the colorants were at least as sensitive as Blue Wool no. 1. However, because the extent of exposure to ultraviolet light was probably rather large and variable, and some numerical details regarding the light sources are unknown, conclusions regarding exposure of these anthraquinone colorants to photoflash cannot be drawn. Additional experimentation may be appropriate.

Red Japanese dyes on cotton

The relative amount of fading or discoloration of cotton fabric dyed with each of three organic red colorants was studied by Kenjo (1987). A master dyer prepared the samples using colorants that had been naturally aged in the dark for twenty-five to fifty years. All samples were irradiated with the same total energy of monochromatic light at each of several wavelengths from 338 to 699 nm. *Sobi* (Japanese alizarin) was the least discolored at all wavelengths, and safflower was the most faded. Sappan wood red, which is related to brazilwood (sappan is a different species of the same genus), was as stable as *sobi* to yellow and red light but was very unstable to blue and ultraviolet light (wavelengths of 498 nm or less). Kenjo's results lead to the conclusion that about 10^3 flashes from a xenon photoflash lamp may represent a maximum allowable exposure for these fugitive natural red colorants (see chap. 3 for details). This limit is similar to that based on the data of Feller, Curran, and Bailie (1984), which were discussed earlier in the section on the light sensitivity of the colorants in Japanese woodblock prints.

> **Conservators' Note:** *The evidence from the work of these investigators again suggests that turmeric and some red natural dyes on textiles are among the most light-sensitive substances considered in this literature search. One hundred flash exposures may exceed the maximum allowable over the total display lifetime of the object. Other red dyes, such as some lichen reds and safflower, some flavonoid derivatives, and logwood mordanted with specific metals, are not quite as light sensitive. They may tolerate a few hundred flashes without suffering a significant decrease in their display lifetimes. Still other natural colorants that are considered moderately light fugitive on textiles can withstand about 10^3 flashes without undue decrease in their display lifetimes.*

Example of display lighting policy for dyed textiles

The Metropolitan Museum of Art had a policy in place in the late 1980s of not exhibiting tapestries containing orchil, brazilwood, or yellow (flavonoid?) natural dyes (Kajitani 1989). It would follow that these objects should not be photographed with flashlamps if this can be avoided. The tapestries that are permitted on display are exposed to less than 80 lux for no more than 60 hours per week if the colorants are present as darker shades. Lighting levels for tapestries with light shades of dyes, even of the more stable indigo and madder, are limited to 30 lux.

The maximum display period allowed is six months; shorter times might be imposed, especially for silk fabrics. Prohibition of gallery flash photography would be consistent with these cautious policies.

Experimental studies on degraded natural dyes on textiles

Some recent research has been directed toward identification and chemical characterization of degraded natural colorants on historic textiles (e.g., Nasu, Nakazawa, and Kashiwagi 1985; Needles, Cassman, and Collins 1986; Quye, Wouters, and Boon 1996). In all of these cases, the exposure times used to age experimental samples have purposely exceeded that necessary to cause a just-perceptible change in the colorants. The results of Nasu and coworkers suggest that a maximum of several thousand photoflashes from handheld cameras or a few hundred from a studio strobe setup might be allowable for several of the yellow flavonoid derivatives they studied (see chap. 3). The relatively high proportion of near-ultraviolet energy in their experimental exposures may make this suggestion overcautious.

Needles, Cassman, and Collins (1986) studied samples of wool and silk dyed with alizarin, brasilin, or carminic acid, with and without various mordants. They irradiated these acidic samples for 80 hours in a weathering device and also buried some samples in sandy loam for two weeks. They monitored not only color, but also tensile properties. The light-induced color changes were significantly affected by the mordant used. The greatest changes were about 1.5 times those observed by Nasu and his group for yellow colorants (at twice the exposure times, which is consistent with the somewhat lower sensitivity of the natural red colorants). In some cases, where the colors did not change much, there were significant reductions in tensile strength, especially for silk samples.

> **Conservators' Note:** *More-specific numerical analyses cannot be performed in the absence of additional experimental details. The range of results, however, suggests that when the identity of the mordant on a historic, red-dyed silk textile is not known, unrestricted flash photography of the object might best be avoided.*

Quye, Wouters, and Boon (1996) have investigated chemical changes in the flavonoids from onion skin, weld, quercitron, old fustic, and dyer's greenweed on wool after one to four weeks of continuous exposure either to a Microscal lamp or to fluorescent lamps filtered to remove all wavelengths below 400 nm. The numerical data they published on the relevant amounts of original compounds and of breakdown products are all for the longest exposure time (672 hours total). These data from a work in progress do not permit calculations of photoflash exposures, but some qualitative information presented by the authors is relevant to the current inquiry. Four samples had a major color component that increased with light exposure; these were extract from onion skin, "quercitin" (probably extracted from quercitron bark), weld, and genistein (a minor colorant extracted from dyer's greenweed). The color change occurred to a greater extent when the samples were exposed to the visible light from fluorescent tubes than when exposed to the combination of visible and near-ultraviolet light from the Microscal lamp. This might possibly be due in part to ultraviolet-induced degradative reactions depleting some of the additional colorant formed by visible-light exposure. Other changes (some loss of major and minor components; the appearance of other ultraviolet-absorbing compounds, either colored, colorless, or both) occurred in all cases. Because of variations in the relative amounts of the colored and uncolored products formed under the different lighting conditions, it was not always possible for the authors to predict whether fading or a color change would predominate. Additional reports from this research project should be of considerable interest. Analyses of kinetics experiments on these types of samples would also be very useful, especially for short or intermittent light exposures.

> **Conservators' Note:** *Studies of these types provide not only data on the dye fading processes but also information useful for determining the original identity of the historic dyes. The investigations also emphasize the fact that natural dyes that are already faded or discolored may undergo further changes upon additional exposure to light. In some cases, degradation products could be responsible for phototendering of the textile support. Additional research in this area should continue to provide data useful to conservators.*

Textiles containing synthetic colorants

Historically, early synthetic colorants of various chemical classes were sometimes introduced as textile dyes before lightfastness information was available (e.g., Allen, McKellar, and Mohajerani 1980; McLaren 1983). Some of these compounds proved to be highly fugitive on fabrics. Over the past century also, newly synthesized colorants occasionally have demonstrated unexpected sensitivity to light on particular textiles. For example, Baumgarte and Wegerle (1986a) noted that azo dyes synthesized in the 1930s sometimes displayed unacceptable photochromic effects on fibers more hydrophobic than cellulose.

> **Conservators' Note:** *Historic objects in pristine condition that are dyed with unidentified early synthetic dyes should ideally not be subjected to any unnecessary light exposure. Flash photography of all kinds should be kept to a minimum for these objects. Early examples of any newly produced dye-fabric combinations might best be displayed and photographed with caution, especially if the identity of the colorant—and its probable lightfastness—is in question. Unrestricted gallery flash photography of these types of objects, which are usually displayed under very carefully controlled lighting conditions, should probably be avoided.*

The physical state of dyes in textiles and the relationship between aggregation and lightfastness have been reviewed by Shore (1990) for textiles dyed commercially with synthetic colorants. Most modern studies of the lightfastness of synthetic dyes on textiles have been performed using exposure times long enough to cause fading that is noticeable to the eye. In most cases, the data have been presented in a comparative manner, that is, with reference to Blue Wool standards, to the equivalent color change on a standardized gray scale, or to both. The light sources used have almost always emitted significant amounts of near-ultraviolet light (for example, sunlight). The literature is extensive; a few references that contain qualitatively or, less frequently, quantitatively useful information are cited here.

Anthraquinone and azo dyes

Allen and McKellar (1980b) reviewed the lightfastness of a large number of anthraquinone dyes on synthetic polymeric fibers. They found some correlations between types and locations of substituents, the nature of the substrate, and lightfastness. Fastnesses ranged from Blue Wool no. 1 to no. 8 (see chap. 3 for an example of data obtained earlier by these authors for a particular dye). It is probably inappropriate to generalize this sort of information, however. Without knowledge of the chemical structures of individual dyes, conservators are likely to find prediction of dye stability difficult. Colorants that might be assumed to be closely related can turn out to have very different fastness ratings.

Rusznák, Frankl, and Gombkötő (1985) investigated the lightfastness of a blue sulfonyl derivative of an anthraquinone dye and two triazinyl derivatives of red azo reactive dyes on wool and cotton; the samples were exposed to 50 or more hours of light from a mercury-arc lamp or a Xenotest apparatus. All the dyes were more sensitive on cotton than on wool. C.I. Reactive Red 2, the monoazo dye, was the most sensitive of the three to the Xenotest light, showing a ΔE greater than 3 in 50 hours.

> **Conservators' Note:** *Calculations based on this result suggest that textiles dyed with the reactive dyes studied could be exposed to a maximum of about 10^5 flashes from a studio xenon lamp over their entire display lifetimes.*

The photochemistry of azo dyes in the presence of various polymeric substances has been reviewed by Griffiths (1980). Blue and red disperse dyes, anthraquinone and azo derivatives, respectively, have been studied by Seves, de Marco, and Siciliano (1995) as colorants for mixed polyolefin yarns. These dyes are known to have reasonable lightfastness (defined with respect to Blue Wool standards) on polyesters and nylon. The blue dye was somewhat less stable on the polyolefin fibers, however, and the red dye, C.I. Disperse Red 13, was found to have a rating equivalent to Blue Wool no. 1 on polypropylene when tested in a Xenotest instrument.

> **Conservators' Note:** *Although specific flash exposures cannot be calculated on the basis of the data provided,*

a reasonable precaution might be to limit flash photography of objects containing polypropylene fibers dyed with C.I. Disperse Red 13.

Hallas (1979) reviewed the photophysics and photochemistry of a variety of disperse azo dyes that contained a 4-aminobenzene substituent; the report includes a discussion of the dyes' lightfastness on several fibers. The author concludes that correlations between dye structure, fiber chemical composition, and lightfastness rating are not straightforward. As is the case for anthraquinone dyes in textiles, minor changes in the chemical structures of the azo dyes studied resulted in some major changes in fastness. No numerical details were presented in the review. Kamel and coworkers (1979) have also shown that slight variations in substituents in the structures of a related group of azo dyes on nylon fabric can lead to wide variations in lightfastness and that the differences are not always predictable. More recently, several authors have gathered evidence leading to similar conclusions for a range of subclasses of azo dyes on several different textile substrates. For example, Marcandalli's group (Massafra et al. 1998) has obtained data suggesting that an orange azo dye on silk would be able to withstand no more than 10^3 flashes from a xenon studio strobe setup (for details on this and other azo dye results, see the corresponding section of chap. 3).

Conservators' Note: *The results of these comparative studies of anthraquinone and azo dyes serve to emphasize the facts that (1) colorants that are structurally similar may have widely varying lightfastness properties, and (2) the response of these colorants to light can be strongly influenced by the nature of the substrate and the presence of additives. These considerations were reviewed by Giles, Duff, and Sinclair (1982), who also considered the roles of dye aggregation in the textile and of hydrogen bonding within the dye molecules and aggregates. In the absence of positive identification of a dye or the specific fastness information on a dyed textile or both, conservators would probably be wise to assume they are dealing with a sensitive example of the particular dye class.*

An example of a less sensitive azo dye on various textiles is reported by Katsuda and coworkers (1996). They studied the photodegradation of C.I. Disperse Red 73 dyed to a medium shade on nylon and polyester, and they also compared their results to those for a reactive red dye on cotton. The samples were exposed to wavelength bands from 200 to 700 nm from a xenon arc that had been dispersed by a large diffraction grating. The investigators observed that wavelengths greater than 370 nm did not have a significant effect on the dyed polyester or cotton samples. The maximum sensitivity of the dyed nylon fabric was to light of this wavelength, however, and blue light was also somewhat effective. One interesting finding here is that 370 nm is the wavelength at which the nylon itself was maximally discolored by light from the xenon arc, suggesting that near-ultraviolet and blue light absorbed by the textile may lead to photodegradation of the dye. Data supplied in the report imply that the quantum yield for fading was somewhat less than 0.001.

> **Conservators' Note:** *If the maximum lifetime allowable change in color due to flash exposure of a nylon textile dyed with this azo colorant corresponds to a* ΔE *of 1, about* 2×10^6 *flashes from a xenon lamp might be allowed. For example, if the object is to be displayed twenty times over its lifetime, then approximately* 10^5 *flashes would be permissible for each exhibit. Only an immensely popular object might possibly be exposed to this many flashes. This conclusion, reached for a relatively light-stable dye-textile combination, stands in contrast to those given previously and in chapter 3 for more-sensitive dyes. The wide range of stabilities to light of various anthraquinone and azo dyes on textiles should be apparent. Appropriate exposure guidelines for photoflash will range from very restrictive (limited studio strobe use) to rather liberal (on the order of* 10^5 *flashes per exhibit).*

Triphenylmethane dyes

Allen, McKellar, and Mohajerani (1980) have assessed the lightfastness with respect to the Blue Wool standards of several common triphenylmethane dyes on acrylic and cotton in a Xenotest apparatus. All of the dyes on cotton were given a rating of Blue Wool no. 1.

> **Conservators' Note:** *With the proviso that the extent of exposure to near-ultraviolet wavelengths may have been higher in the aging apparatus than it would be from photoflash, it can be suggested that several thousand flashes per exhibit over the display lifetime of an object may be the maximum allowable for cotton dyed with such triphenylmethane dyes as Brilliant Green YN, crystal violet, or Basic Red 9 (pararosaniline).*

Effects of relative humidity

Giles, Haslam, and Duff (1976) obtained experimental evidence for the occurrence of oxidative or reductive processes (or both) involved in the fading of several sulfonated acid dyes on cotton or wool. They found that lightfastness decreased as relative humidity increased for the dyes on cotton. The effect was not significant for the dyed wool samples. The authors suggested that their results were evidence for oxidative degradation processes in the dyed cotton and that the increase in water in the textile with increasing relative humidity facilitated the transport of oxygen to the dye molecules. The most sensitive dyes were C.I. Direct Red 2 (an azo dye) at a 0.1% dyeing on cotton, and C.I. Acid Blue 15 at 1% on wool. These colorants had a sensitivity equivalent to Blue Wool no. 1 to light from a Microscal lamp at relative humidities of 45% or higher. C.I. Direct Blue 1 at a 1% dyeing on wool was equally sensitive at low relative humidities.

> **Conservators' Note:** *Restriction of gallery flash photography is suggested by these findings, especially if textiles dyed with these colorants are in a very popular display; if they are on display in a building where the typical RH of 50 ± 5% is maintained; or if they will be subjected to higher humidities.*

Other dyes on textiles

Oger (1996) investigated the effectiveness of fixatives on the fastness of several Ciba-Geigy direct dyes that are used by French conservators on linen lining fabrics and threads. The most sensitive dye was Orange TGL; the data suggest that this orange dye could be allowed a maximum of about 2×10^4 flashes from a xenon strobe without color correction during its display lifetime (see chap. 3). The other dyes investigated were from 0.1 to 0.5 as sensitive. Of more interest, the unfixed samples were all somewhat less fugitive to light than were the corresponding fixed samples.

Wilkinson, McGarvey, and Worrall (1992) applied the technique of diffuse reflectance flash photolysis to dyed textiles and other opaque polymers to detect transient species, thereby increasing understanding of photochemical reaction mechanisms and kinetics. They studied the halogenated fluorescein derivative rose bengal on nylon and cotton, and sulfonated aluminum phthalocyanine on cotton, observing that about 2% of the dye molecules did not recover from a flash exposure (Wilkinson, Willsher, and Pritchard 1985; Wilkinson and Willsher 1987). This result suggested a quantum yield for photodegradation on the order of 0.01 (see chap. 3 for experimental details and calculations).

> **Conservators' Note:** *Calculations based on the data provided in these reports suggest that gallery flash photography of nylon dyed with rose bengal should probably not be allowed, and studio photography should be monitored and documented.*

Phototendering of dyed cotton, polyamide fibers, and wool

An historical perspective on the study of phototendering has been presented by McLaren (1983). Early research was generally done using long exposures to sunlight or mercury-arc light sources that contained large amounts of ultraviolet-B as well as ultraviolet-A (e.g., Egerton 1948), so these older results are not useful for the present purpose. Egerton also presented evidence at this early date that the white pigments titanium dioxide and zinc oxide accelerated photodegradation of textiles (see "Fluorescent Whitening Agents," pages 102–8). Phillips and Arthur (1985) have reviewed the photochemistry of the sensitization of cellulose photodegradation by dyes. Although they identify anthraquinone sulfonates specifically as effective sensitizers, they do not provide any numerical data on yields or on the effectiveness of visible light. Baumgarte and Wegerle (1986b) presented a detailed review of catalytic fading of dye mixtures and of phototendering in cellulose by anthraquinone and vat dyes, but they did not include enough experimental details to permit calculations. Allen (1989) presented a concise review of the probable mechanisms of phototendering, using examples of several different types of dyes that cause this phenomenon.

Rånby (1989) showed that the rate of degradation of cotton is increased sixfold when the cotton is dyed with a yellow anthraquinone derivative and that these rates are higher still when the relative humidity is high. The data presented earlier in the section on cotton suggest, however, that even a tenfold increase in the rate of photodegradation of cotton would not be enough to cause unacceptable change due to moderate amounts of xenon photoflash exposure. The photostability of the dye itself may be lower than that of the substrate in this case.

The effect of five triphenylmethane dyes on the photodegradation of a cotton substrate was investigated by Ladisch, Brown, and Showell (1983), who aged their samples in a xenon-arc fading apparatus. The minimum change in appearance used in the experiments was equivalent to Blue Wool no. 5. Four of the five dyes provided some protection to the fabric, as measured by breaking strength, but C.I. Reactive Blue 53 appeared to cause phototendering after very long exposures. The data do not permit calculations useful to the present discussion.

The photostability of dyed polyamide fibers was investigated by Reinert (1988), who described this fiber as "having the poorest light stability of all synthetic fibers" at that time. Using polyamide fibers dyed to different depths of shade with a beige-to-brown dye mixture or with the yellow dye C.I. Acid Yellow 129, the author demonstrated that the tear strengths of the dyed yarns could be reduced by up to 90% compared to those of the undyed fibers. Long exposures to sources that contained rather large amounts of near-ultraviolet light were used in these studies. The extent of the effect depended on the time of exposure, the precise ultraviolet content of the radiation, and the relative humidity (see chap. 3). The results indicated that, in general, moderate amounts of flash photography would probably not cause unacceptable weakening of the fabric.

In contrast to these results on phototendering of cotton and polyamide fibers, Rusznák, Frankl, and Gombkötő (1985) have shown that reactive dyes can retard the photodegradation of wool. From four to eight times as much radiation was required to cause the same increase in alkali soluble matter in the dyed as in the undyed fibers for an anthraquinone, an azo, and a diazo reactive dye. Unacceptable fading of the dyes occurred long before the wool fibers showed deleterious weakening.

> **Conservators' Note:** *These evaluations of phototendering suggest that the dyes usually increase the rate of degradation of the textile support by a factor well under 10, compared to the rate of degradation of the undyed textile. Often the sensitizing dyes themselves will degrade more quickly than the fibers. Thus, the occurrence of the phototendering phenomenon will usually not have a*

significant effect on the order-of-magnitude approximations of the amount of allowable photoflash exposure for the dyed fibers.

Organic dyes and pigments in other polymeric materials

Many photochemical studies of colorants have been performed on dye samples included in a clear, cast film of polymeric material, such as cellulose or acrylic. Before diffuse reflectance methods were developed, this was the method of choice to investigate the response to light of colorants in a solid matrix. Included in this section are some evaluations of data obtained from these types of studies, as well as discussions of relevant information on other dye–polymer systems, such as colored paper or opaque plastics. Also of interest is the possible photosensitization (or, conversely, the protection from photochemical degradation) of the polymeric material by the colorant. Allen and Edge (1992d) and Klemchuk (1983) have addressed these latter concerns in reviews.

A sampling of the large variety of investigations that have been conducted on specific synthetic dyes in particular polymers is gathered in the corresponding section in chapter 3. The range of light sensitivities represented is very large, from a maximum number of photoflashes over display lifetimes of under 100 to well over 10^6. Lightfastness of dyes in these materials depends not only on the colorant, the substrate, and the wavelength of the light but also on the environmental conditions; the presence of other materials, such as stabilizers, colorant combinations, or polymerization aids; the aggregation state of the dye in the polymer; and so on. Some of the most sensitive combinations are mentioned here. Other results that may be relevant to risk assessment of light exposures are also presented.

Dyes in acrylic, epoxy, or polyester resins

Tennent and Townsend (1984) monitored the lightfastness of several different solvent dyes and two pigments when the colorants were included in acrylic, epoxy, or polyester resins. Almost all of the dyes were most stable in the acrylic polymer. The hardener in the epoxy may have contributed significantly to dye fading (see summary of pertinent details in chap. 3).

In a study of the lightfastness of epoxy resins suitable for mending stained glass, Tennent (1979) measured the fastness of polymer samples containing a wide variety of colorants including azo, quinophthalone, indanthrone, methine, and phthalocyanine dyes as well as carbon black. Because these colorants had been chosen to withstand exposure to sunlight in a stained-glass window, they can be expected to be rather stable to radiation from testing apparatus. The results confirmed this; the least stable dye was C.I. Solvent Red 7, a chromium complex of an azo dye, which had a lightfastness equivalent to Blue Wool nos. 6 and 7 when exposed to the Microscal 500 W mercury–tungsten lamp for 1000 hours. This exposure is very roughly equivalent to more than 10^8 flashes from a studio xenon strobe lamp without near-ultraviolet filtering. It can be concluded that the particular colorants used in this study would not be in danger of fading from exposure to reasonable amounts of flash photography.

Allen and coworkers (1983) studied the photodegradation of three anthraquinone dyes included in epoxy resin films as a function of time and of the presence of various hardeners and stabilizers. The overall quantum yield may have been as high as about 0.05, which is 5000 times higher than the value assumed for the calculations of the effect of flashlamps (see "Xenon lamps," pages 15–20; for details of the study, see chap. 3). These authors also observed that the stabilizers Tinuven 770 and tetracyanoethylene, although initially somewhat protective, lost their protective capacity after several hours of irradiation. At this point, the rate of fading of the anthraquinone dyes in the epoxy films increased markedly. The behavior suggests that the stabilizers acted in a sacrificial rather than a catalytic manner. Some stabilizers that were capable of acting as traps for free radicals were more effective; in fact, 4-hydroxy-2,2,6,6-tetramethylpiperidino-*N*-oxy essentially prevented fading of the dyes over the course of the experiment.

> **Conservators' Note:** *The results obtained by Allen and his colleagues suggest that the particular anthraquinone dyes tested—if used to color epoxy mends or fills or in coatings on artifacts—may be sensitive to a moderate amount of exposure to flashlamps containing a small near-ultraviolet component. Additionally, anthraquinone dyes in epoxy resin formulations that do not contain stabilizers to scavenge free radicals may fade after they have received moderate light exposures. The response to repeated photoflash of some anthraquinone dyes in such multicomponent systems might profitably be further investigated.*

Azo and triphenylmethane dyes in hydrophobic polymers

Giles and coworkers (1974) investigated the role of temperature and relative humidity in the fading of a variety of disperse azo and triphenylmethane dyes in hydrophobic polymers. The temperatures used were 15 °C and 60–102 °C. Increasing the temperature increased the fading for the dyes in cellulose acetate, nylon, and acrylic. The extent of the effect appeared to depend on the strength of the dye–polymer interaction. Increasing relative humidity also increased the rate of dye fading in nylon and in cellulose acetate, especially at elevated temperatures. Relative humidity did not appear to have a significant effect on the dyes in acrylic. Further investigation of this phenomenon was pursued using films of gelatin and anodized aluminum, as well as cellulose acetate and textile substrates (Giles, Haslam, and Duff 1976). The extent of the relative-humidity effect depended on whether the photodegradation processes were oxidative. If oxygen is involved in the breakdown of the dye and increasing the water content of the support can enhance the access of oxygen to the dye molecules, then increasing relative humidity is likely to increase the rate of dye fading. Additional research by Giles and colleagues (Giles, Hojiwala, and Shah 1972; Giles et al. 1972) is summarized in chapter 3.

> **Conservators' Note:** *The data suggest that the most sensitive dye–polymer combinations tested by Giles and his colleagues would have a lifetime flash exposure limit between 10^3 and 10^4 flashes.*

Azo and anthraquinone dyes in films of ethyl cellulose, cellulose diacetate, and polyamide

Rembold and Kramer (1978) investigated the catalytic fading of some azo and anthraquinone dye mixtures in ethyl cellulose films under various irradiation conditions. For example, when an anthraquinone dye that absorbed light of wavelengths greater than 500 nm and a yellow azo dye that absorbed only shorter wavelengths were present together in a film, the yellow dye faded when the sample was exposed only to wavelengths above 500 nm. The worst-case results reported, when only visible light was used as the source, were about 30% fading of C.I. Acid Yellow 29 or C.I. Acid Yellow 127 in the presence of C.I. Acid Violet 47 or C.I. Disperse Blue 56. If near-ultraviolet wavelengths were included in the radiation, all of the dyes, individually or in combinations, faded at least as much as the worst-case combination described here.

> **Conservators' Note:** *The data presented for these samples suggest that the acceptable total change due to flash would be reached by about 5×10^4 flashes of a xenon lamp with ultraviolet blocked. However, at least an order of magnitude fewer flashes from a source including near-ultraviolet light would exceed the acceptable lifetime limit for these colorant combinations.*

Tera and coworkers (1985a, 1985b) studied the fading of seven different 4-hydroxyazobenzene dyes in cellulose diacetate and polyamide films. Lightfastness depended markedly on the extent of aggregation of the dye in the polymer, on the nature of the polymer, and on the nature and location of the substituent on the benzene moiety in the dye molecule. The most-sensitive combinations investigated have a lifetime permissible flash limit of roughly 2×10^3 flashes from a studio strobe without color correction (see chap. 3). Tera and colleagues (1985b) also noted, however, that the rate of fading of this dye decreased rapidly after the first 60 hours of exposure; that is, the dye appears not to have obeyed the reciprocity principle.

> **Conservators' Note:** *This observation serves to emphasize the importance of taking special precautions with dyed polymeric objects that are in pristine condition. In contrast, objects containing these dyes in an already faded state are not likely to continue changing at the same rate when they are exposed to more visible light or more near-ultraviolet light.*

Vinylsulfonyl derivatives of reactive dyes in cellophane films
Okada and his colleagues (Okada, Kato, et al. 1990; Okada, Hirose, et al. 1990a, 1990b) have investigated the properties of vinylsulfonyl derivatives of a variety of reactive dyes. The dyes were included singly or in binary mixtures in cellophane films and irradiated in a carbon-arc apparatus at a range of relative humidities. Although the light source was usually not filtered, occasionally cutoff filters were employed (data obtained under these conditions are evaluated in chap. 3). Some of the dyes, under certain conditions, appear to be extremely sensitive to light; for example, a reactive vinylsulfonyl azo dye at very high humidities in the presence of dye sensitizers may be able to tolerate only tens of flashes during its entire display lifetime. This research group continues to investigate the photofading behavior under humid conditions of mixtures of vinylsulfonyl dye derivatives in cellophane films. The results are complex and do not appear to be predictable on the basis of the type of dye or the nature of the substituents.

Dyes in poly(vinyl acetate) glazes and in poly(vinyl alcohol) films
An extensive investigation of the fading of transparent poly(vinyl acetate) glazes containing the azo colorant C.I. Pigment Red 66 (C.I. 18000:1) is reported by Whitmore and Bailie (1997). This material exhibits the uncommon fading behavior in which the rate of change in appearance increases with time after a long exposure during which little change takes place. In cleverly designed experiments these investigators show that the kinetics of appearance changes are not the same as the kinetics of photochemical modification of the colorant in the glazes. At some pigment concentrations the glazes could withstand a few million lux hours of exposure to visible light from a xenon arc with only a minor change in appearance. Subsequently, a relatively small additional exposure (judged from the published figures to be less than 10^5 lux hours) would result in significant changes in appearance. In other words, whether or not the pigment and the polymer individually obey the reciprocity principle, the glaze formulation does not. The researchers modeled the glaze-fading behavior mathematically by considering the kinetics of changes in the optical properties of the glaze as the dye molecules are chemically altered. They convincingly show that, although the pigment is lost steadily over time, changes in the appearance of the colored glaze do not occur in a gradual manner. Also, glazes of intermediate chroma (as distinct from highly colored or pale glazes) are the most likely to show abrupter changes in appearance.

> **Conservators' Note:** *These findings emphasize the importance of keeping records of light exposures for objects that may be light sensitive. Exposure to photoflash sources, with ultraviolet blocked, would not be deleterious unless an object with a C.I. Pigment Red 66 glaze already had an extensive display history. In their article, the authors offer very good suggestions for monitoring and display of objects with vulnerable clear glazes.*

Allen, Mohajerani, and Richards (1981) studied the effects of a radical scavenger and an electron trap on the photofading of basic triphenylmethane dyes in poly(vinyl alcohol) films. The data they reported for the dye malachite green in the absence of compounds that block electron transfer suggest that the quantum yield for fading of this dye is greater than 0.001 in the solid state. This is a hundred times more sensitive than the yield assumed in the example calculations described in chapter 1 (see under "Light Sources," pages 13–25) and could imply a lifetime limit for flash exposure of several hundred to a few thousand flashes for colorants at low concentration in this polymer. Pararosaniline, the other triphenylmethane dye used by Allen and coworkers in this study, appears to be more than an order of magnitude less sensitive in poly(vinyl alcohol) films.

Other dyes in film-forming polymers

Additional investigations on the lightfastness of the common classes of dyes in polymeric films have been undertaken by several groups of researchers. Zweig and Henderson (1975) obtained some useful data while studying the effect of a singlet oxygen quencher on the fading of a variety of dyes. Anthraquinone derivatives have been investigated, for example, by McAlpine and Sinclair (1977); Zamotaev, Mityukhin, and Luzgarev (1992); and Lishan and coworkers (1988). The latter group, as well as Ball and Nicholls (1984), have studied azo dyes; Jones, Oh, and Goswami (1991) and Duxbury (1994a, 1994b) have done work on triphenylmethane dyes; and Matsushima, Mizuno, and Itoh (1995) have investigated model compounds for flavylium dyes in poly(methacrylic acid) films (see corresponding section of chap. 3 for details on these studies).

> **Conservators' Note:** *This sampling of the literature demonstrates the wide range of sensitivities to light of dyes in various solid polymeric media. The allowable exposures to photoflash depend on several criteria: the nature of the colorant; the colorant's chemical form and physical state in the particular substrate; the environmental conditions; the presence of a sensitizer; the prior storage and display history of the object; and so on. A cautious recommendation for approximate number of flashes can range from a few hundred to more than* 10^6.

Other specific dye–polymer systems

Numerous studies of other types of dye–polymer systems may be found in the literature. Selected examples are summarized here and in chapter 3. Although these investigations may have been undertaken with an unusual application in mind or employ uncommon materials as samples, the information generated may be useful in the context of the effects of light on art and archival collections.

Tennent, Townsend, and Davis (1982) studied the possibility of using phenothiazine in a poly(vinyl chloride) film as an ultraviolet dosimeter. They subjected samples to various types of gallery illumination as well as to fading in a Microscal apparatus with and without cutoff filters of plastic glazing materials. From their data (see chap. 3) it can be calculated that an approximate allowable exposure to xenon photo-flash without complete ultraviolet blocking is a few thousand flashes for a studio strobe and approximately an order of magnitude more for a point-and-shoot camera. The authors also noted that in the absence of filters, the decrease in transmission (i.e., the coloration of the dye) was eventually reversed, and the colored product was bleached by long exposure to near-ultraviolet and visible light.

The slight photochemical degradation of rose bengal, which was reported by Wilkinson, Willsher, and Pritchard (1985) and described earlier (see "Other dyes on textiles," page 88) was also observed when the dye was bound to beads of copolymers—either chloromethylated styrene-divinyl copolymer or the polymer obtained by copolymerization of chloromethylstyrene and the mono-methylacrylate ester of ethylene glycol. This result suggests that photodegradation of the dye occurs via mechanisms that are not dependent on chemical interactions between the dye and specific fibrous substrates. The data given for the dye on the polymer beads lead to the same numerical results as did the calculations described earlier for rose bengal on cotton fibers.

Dubois and coworkers (1996) compared the quantum efficiencies for photochemical change of several different types of dyes in solution and in solid matrices. Among the colorants studied were some rhodamine dyes. The researchers found that although the quantum yields for photochemical degradation of these compounds were about 0.01 in solution, when the colorants were embedded in the clear plastic matrices the yields fell by about five orders of magnitude to about 10^{-7}.

> **Conservators' Note:** *These results suggest that xanthine dyes of this type, when used as colorants in plastics, are probably not likely to be affected by moderate exposure to photographic and reprographic light sources. The authors comment, however, that a large proportion of the apparent protection of the dyes in the solid matrix may be due to the fact that oxygen cannot reach the colorants in this state. Materials that are porous to air may not be able to afford the same extent of protection to these dyes.*

Jones, Feng, and Oh (1995) and Fisher, Lewis, and Madill (1976) observed the photochemical behavior of examples of other xanthine and related dyes when they were associated with proteinaceous polymers; and Seybold and Wagenblast (1989) tested the outdoor photochemical stability of perylene and violanthrone (see chap. 3 for summaries of results). None of these studies suggested light sensitivities that should be of concern for collections care.

Organic colorants in mixed systems

Conservators often approach the care of problematic objects containing mixed media of unknown or complicated (or both) components by analyzing the materials present and by testing treatment procedures on carefully chosen or prepared models. These mixed systems have frequently included colorants. A few studies of mixed systems containing organic dyes or pigments and polymeric materials that may be found in art and archival objects are considered here.

Moura, Oliveira-Campos, and Griffiths (1997) review the effects of various additives on the photostability of dyed polymers. Substances such as ultraviolet absorbers, singlet oxygen quenchers, and free-radical scavengers, which are intended to increase the stability of the materials, are included in their survey. They describe some mechanisms of action, but the information provided on lightfastness is mostly qualitative. Many of the original sources cited are in the patent literature, and the authors do not provide the experimental details necessary for calculating photoflash exposures.

The lightfastness of handmade papers containing only cotton fibers, organic colorants, and a commercial fixative was investigated by Levison, Sutil, and Vanderbrink (1987). A variety of industrial organic colorants and some inorganic pigments were used. Because the samples were exposed to sunlight, sunlight filtered by window glass, or fluorescent light until Blue Wool no. 7 showed a change, the data cannot be used to determine allowable photoflash exposures. The observation that the stability of the pigmented papers ranged from excellent to unsatisfactory is consistent with the results of other investigations.

Ellis and Yeh (1997) summarize information on wax-based drawing media; these range from traditional materials, such as oil pastels, to modern, unconventional media, such as lipstick. The authors list the methods of manufacture and the typical ingredients in these mixtures and provide qualitative comments on light stability when it is known. Numerical data that could be used in the present evaluation are not included, but the background information provided should be useful if lightfastness studies are undertaken.

Inks on paper

Concerns have been raised regarding the lightfastness of inks in commercial fiber-tipped pens, most recently in conjunction with the tendency of these media to transfer to other objects with which they are in contact (Ellis 1994; Sutherland 1994). Ellis was able to bleach a transferred purplish color by exposing the paper that contained the migrated material to daylight filtered through window glass. Complete bleaching required less than three weeks. Sutherland noted that the lightfastness of these inks on objects already in collections is generally unpredictable.

Lafontaine (1979) investigated the lightfastness of almost 550 commercial felt-tip pens available in the late 1970s and classified his results as good, tolerable, or unacceptable. Color was not a predictor of stability. The experimental details provided and the assumptions about the light sources used permit the calculation that roughly 10^4 xenon studio strobe flashes with near-ultraviolet light blocked would be the maximum tolerable by the least stable inks on acid-free papers (see chap. 3). Coverage by the inks is not specified in the report. These numbers would only apply to felt-tip pen drawings in which the depth of shade is similar to that of the experimental samples.

In an investigation possibly relevant to inks on paper, Kaneko and Yamada (1981) studied the photoreduction of the dye methyl viologen on cellulosic paper. This compound, a chloride salt of a dipyridyl derivative, has been used extensively in oxidation–reduction research. The authors used what they called white light, which included visible and near-ultraviolet wavelengths; visible light from 400 to 800 nm; or monochromatic 460 nm light isolated from a halogen lamp by filters. A calculation based on the experimental data supplied and on the assumption of a relatively low dye coverage on paper suggests that this redox dye could be subjected to about 10^4 flashes from a studio strobe or to several hundred photocopies during the lifetime of the object (see chap. 3).

Analyses of traditional inks on Western and Near Eastern manuscripts have become highly sophisticated (e.g., Sistach and Espadaler 1993). The most common type of ink used for manuscripts and drawings up to the late nineteenth century was iron gall ink (Perkinson 1974). This class of inks is usually a purply black initially, but the inks have often faded to brown because of extensive light exposure over time. Although Perkinson mentions that additional light exposure, including near ultraviolet, will fade them further, he does not provide any numerical data or references. Both original and degraded components can now be identified and quantitated. The effects of light exposures under controlled conditions on well-documented samples have not been reported, however. For data about the tannin colorants in the inks, a survey of the technical literature from the beer industry might be appropriate, since manufacturers are concerned about the shelf life of their tannin-containing products under conditions that can include light exposure.

Tannins in leather

Methods for investigating the stability of tannins in leathers are also being pursued. Wouters (1993) recently reported analyses of historic leathers and of new and artificially aged leather samples. Several extractable tannin compounds were found in historic leathers, but the only one that absorbs visible light is ellagic acid. The other components do not absorb above 350 nm. Experimental samples that had been subjected to ten years of exposure to room lights plus daylight through window glass displayed some compositional differences compared to dark controls. This observation cannot be used numerically, however, because no controlled light-exposure studies were performed in the investigation.

A study group set up by the European Commission for Protection and Conservation of European Cultural Heritage is evaluating methods for artificially aging leathers (Larsen, Vest, and Kejser 1994). Members of the research group have used light in combination with gaseous pollutants to age samples. The leathers were exposed to ten days of 2×10^4 lux from an unspecified light source. Numerical evaluation of the results with respect to flash exposure is not possible on the basis of the information available. The study participants comment that more research on the effects of light on leathers needs to be done.

Inorganic pigments associated with synthetic polymeric materials

Reviews of studies of the lightfastness of inorganic colorants in synthetic polymers and of photostabilization and photosensitization by these colorants have been published by Klemchuk (1983) and Allen (1989). The former author summarized earlier work showing that, in general, inorganic pigments appeared to confer increased resistance to degradation on low-density polyethylene, unstabilized polypropylene, and poly(vinyl chloride). Carbon black was among the best protectors. Red iron oxide appeared to slightly increase the photosensitivity of thin polypropylene tapes in the presence of some stabilizers as well as a sensitizer. Because the light exposures were very high in these experiments and the polymers are not normally affected by the wavelengths emitted by xenon flashlamps, the observed increase in sensitivity is probably not cause for concern. In contrast, Allen (1989; Allen and Edge 1992b) suggested that certain inorganic colorants such as cadmium yellow, ultramarine blue, and C.I. Permanent Red 2B might act as photosensitizers in certain applications; they recommended that research be performed in this area.

Colored inorganic pigments in various polymers

Ogbobe and Ossai (1992) investigated the change in tensile properties and solubility of high-density polyethylene films containing 1% by weight of commercial inorganic pigment formulations of pantonene green and ultramarine blue. The films had been exposed outdoors to day–night cycles at tropical latitudes. Each colorant consisted of a mixture of several pigments and other components, including organic materials. Both pigment formulations decreased the tensile strength of the films, although the differences lessened after several weeks of exposure. The green pigment increased the pliability of the plastic. Because no measures had been taken to eliminate ultraviolet-B wavelengths from the tropical sunlight, the numerical results are not useful to the present inquiry. However, the technique used in the study of casting films from hot solutions (suspensions) of commercial pigments and plastics might be useful for preparation of samples for future lightfastness studies.

Banik and Ponahlo (1983) investigated the effects of exposure to light and pollutants on mixtures of pulped, hundred-year-old willow paper and copper green pigments. The samples were exposed for 75 hours to an Austrian quartz lamp (assumed to be a tungsten–halogen lamp) with wavelengths less than or equal to 340 nm filtered out. The samples were analyzed by infrared spectroscopy. This technique is not likely to detect changes of less than 5% in the chemical composition of the samples. Pulps pigmented with malachite green did not show significant changes after exposure to light. The authors state that similar results were obtained with other pigments tested (various verdigris and copper acetate preparations and copper chloride). On this basis, it can be suggested that these pigments in naturally aged willow pulp or papers might be exposed to about 10^4 flashes from xenon flashlamps without undergoing an easily detectable photochemical change.

Under appropriate conditions, inorganic colorants that also are semiconductors, such as cadmium sulfide, CdS, and zinc sulfide, ZnS, may catalyze photochemical reactions. Darkening of lithopone (a white pigment consisting of zinc sulfide and barium sulfate, $BaSO_4$) and the photoinduced corrosion of the substrate by the lithopone breakdown products zinc sulfate, $ZnSO_4$, and zinc hydroxide, $Zn(OH)_2$, is an example of this type of phenomenon (Kisch and Künneth 1991). Research on the reactions involved is usually performed with colloids of finely divided semiconductor powders in the presence of moisture or solvents.

Watanabe, Takizawa, and Honda (1977), using aqueous slurries of CdS with rhodamine derivatives, observe that quantum yields for photoinduced modifications of the dyes depend on the concentration as well as on the physical state of the CdS and the dye. For example, they report yields of about 0.5 at 600 nm—where the dye itself absorbs—when a large excess of semiconductor is present. Tokumaru and coworkers (1985) report photocatalyzed oxidations approaching 100% for some olefins in acetonitrile-methanol mixtures in the presence of CdS after 6 or 7 hours of irradiation with visible light above 430 nm.

> **Conservators' Note:** *The relevance of these types of findings to conservators' concerns and to photographic flash exposure is not easy to assess because the extent of reaction possible in dry, solid museum objects is not known. However, in the unusual circumstance in which an object containing semiconductor-type inorganic pigments mixed with oxidizable organic colorants or polymers becomes wet or saturated with solvent, it may be wise to avoid all unnecessary exposure to light, including high-intensity flash sources.*

White inorganic pigments in various polymers

Inorganic white pigments are used both as the principal colorant and as fillers and opacifiers in paint and coating formulations. Extensive literature on the durability of these products exists. A few examples of studies from the conservation viewpoint, or those reporting numerical results useful for calculations in the current context, are summarized here (see "Fluorescent Whitening Agents," pages 102–8, for more information on these materials).

White latex paints containing zinc oxide may chalk when exposed to daylight under conditions in which the resin without the pigment does not do this. An early investigation of the chalking phenomenon suggested that the zinc oxide pigment, when irradiated, may promote decomposition of lower-molecular-weight polymers in the paint film and that this occurs despite the coating being cured before exposure (Hoffmann and Saracz 1969). The rutile form of titanium dioxide partially protected against the effect, apparently by screening ultraviolet and visible light. The samples were subjected to regimens of light–dark cycling and to constant exposures at different humidities in a fading apparatus during the study. The extent of chalking was greatest during the initial light exposure. Insufficient experimental details were provided to calculate actual exposures used.

Allen (1989) reviewed the photosensitized degradation of polyalkenes by white pigments such as titanium and zinc oxides. He reports that the rutile form of titanium dioxide has a protective effect on polyethylene, whereas the anatase form acts as a photosensitizer. Polyethylene containing 2% anatase was irradiated in a Xenotest apparatus, and oxidation was monitored as the increase in carbonyl groups. The results suggest that this pigmented polymer could tolerate thousands of flashes from a xenon strobe source without near-ultraviolet filtering. This was the most sensitive sample tested. Earlier work (Allen, Gardette, and Lemaire 1982) had shown that both hindered amine light stabilizers and antioxidants protect against near-ultraviolet-induced degradation of polypropylene containing either anatase or rutile titanium dioxide. The extent of protection depends on the formulation and on the wavelength distribution of the near-ultraviolet light. The data suggest that even without the protecting agents, polypropylene containing titanium dioxide could withstand at least 10^5 flashes of a xenon studio strobe. Allen points out that other white pigments, such as zinc oxide; magnesium oxide, MgO; calcium carbonate; and barium sulfate, act as stabilizers in polyalkenes, although some of them may be effective simply because they reflect a large percentage of the incident light between 300 and 400 nm. Quantitatively, the stabilizing effect will depend on the manufacturing details of the pigment and the polymer and on the chemical nature of the polymeric material.

Ohtani and coworkers (1992) studied the photodegradation of polyethylene films incorporating titanium dioxide powders. Chalking was not observed when wavelengths less than 460 nm were removed from the mercury-arc radiation used in the experiments. The researchers did not directly investigate the effects of light of wavelengths between 400 and 460 nm. They observed rapid and extensive chalking when the samples were irradiated with light of all wavelengths greater than 280 nm. Because the relative effectiveness of radiation of wavelengths between 360 and 460 nm is not specified, calculations cannot be performed on the published data.

Whitmore and Bailie (1990) examined the chalking phenomenon from the point of view of the conservator. The carrier in their studies was a poly(*n*-butyl methacrylate) resin (Elvacite 2044) that did not chalk without the additives. They measured the decrease in solubility and the extent of cross-linking of the resin alone and with various white pigments added, as a function of exposure to near-ultraviolet light from fluorescent lamps at constant temperature and humidity. The presence of two forms of zinc oxide, or of either anatase or rutile titanium dioxide, delayed the onset of cross-linking. Barium sulfate, however, accelerated the cross-linking. Rutile titanium dioxide and the zinc oxide pigment called French Green Seal protected against chalking, but the other additives did not (see chap. 3).

Conservators' Note: *Calculations based on the authors' published data suggest that Elvacite 2044 pigmented with titanium dioxide, zinc oxide, or barium sulfate could*

> *withstand about 10^6 flashes froms studio xenon flash-lamps (see chap.3). The identity of the white pigment in a latex paint should be known, however, before the decision is made to place an object so painted on long-term display in daylight.*

Heller and coworkers (1987) performed a detailed study of the mechanism by which rutile titanium dioxide may participate in photodegradation of paints and coatings from the point of view of the semiconductor properties of the pigment. They state that absorption of light of wavelengths less than or equal to 355 nm by titanium dioxide is necessary to initiate the degradative processes. On this basis, xenon photoflash lamps would not be expected to be harmful to paint containing this pigment.

This conclusion disagrees with earlier results of Egerton and King (1979), who reported on the quantum efficiencies of titanium dioxide–catalyzed oxidation of isopropanol, a system claimed to be an appropriate model for the breakdown of paints containing this white pigment. The researchers observed quantum efficiencies from 0.03 to 0.3, depending on the light intensity, when they irradiated the sample suspensions with blue light (the 404 nm wavelength band isolated from a mercury arc by a narrow band-pass filter). The higher efficiency was only observed at low irradiances. The applicability of these results to coating formulations containing titanium dioxide would depend on the accessibility of hydroxyl groups to photochemically active species.

Pigment-coated papers

The coatings on modern pigment-coated papers are formulations based on polymeric materials. White coatings usually contain one or more inorganic white pigments. It is thus appropriate to consider these composite materials here (see also "Fluorescent Whitening Agents," pages 102–8).

Mizrachi (1994) has presented a detailed analysis of several types of pigment-coated papers in which the pigment is an inorganic white substance. The pigments detected were calcium carbonate, kaolin (aluminum and silicon oxides), titanium dioxide, talc (magnesium and silicon oxides), barium sulfate, sulfo-aluminates, and a polystyrene-based plastic pigment. The amounts of other components are also reported. The author did not investigate the effects of light exposure on these coated papers, but the detailed information on the composition of the pigment coatings will be helpful to such studies in the future.

Mailly and coworkers (1997) have investigated the effects of daylight radiation on bleached chemical pulp papers coated with a styrene-butadiene-acrylic acid latex mixture and on films of the coating materials. Other additives, for example, white pigments ($CaCO_3$ or kaolin), organic fluorescent whitening agents, poly(vinyl alcohol), and soaps or other dispersants were included. A Xenotest xenon lamp with or without an ultraviolet-blocking filter was used as the light source. Some samples were exposed to gaseous pollutants (NO_2 or NH_3). The latex films themselves were initially bleached by the xenon lamp, whether or not wavelengths less than 400 nm were excluded. This response was followed by a significant yellowing only when the light included near-ultraviolet wavelengths. More than 100 hours of irradiation, with near-ultraviolet light included, was required for the yellowing to become significant, which suggests that exposure to more than 10^4 flashes from a strobe lamp without ultraviolet blocking would be required to have a similar effect on a latex film.

The papers coated with the latex yellowed more rapidly than did the latex films themselves or the uncoated papers, even when ultraviolet light was excluded. The initial response in the absence of ultraviolet light was a slight bleaching. Changes in the Yellowness Index that were equivalent to a change detectable by eye had occurred within 24 hours, and NO_2 accelerated the yellowing somewhat. If it is assumed that this apparatus operates under conditions similar to those of the Atlas Weather-Ometer, roughly 10^{22} photons of visible light caused a change detectable by eye. Over the display lifetime of this coated paper, flash exposure should deliver no more than 1% of this, or 10^{20} photons of visible light. Therefore, up to 10^5 flashes from a studio strobe setup would be permissible. Only about two thousand photocopies of the coated paper could safely be made.

Antioxidants appeared to increase yellowing somewhat, up to a certain ratio of antioxidant to latex. The effect is greater when ultraviolet light is included and may depend on the composition of the coating. The effects of ultraviolet photography or extensive photocopying on coated pulp papers with organic antioxidants might be worthy of investigation.

Coatings that have been polymerized in situ by exposure to light will have been fabricated from mixtures of monomers, including photoinitiators or photosensitizers that absorb light and trigger the polymerization reactions. In the past, most of these compounds absorbed ultraviolet light, but recent developments have promoted the use of photoinitiators that absorb in the visible region (Allen 1996). If the photoinitiator is not completely consumed by the polymerization reactions, additional exposure of an object containing the polymer might trigger further photochemistry.

> **Conservators' Note:** *If a pigmented coating on paper has been photopolymerized in the presence of the colorant and the process has gone to completion, the risk from exposure to photoflash may not be significant. In contrast, if residual initiator remains in the coating or if the dye and other components have been added after polymerization, photodegradation of the colorant and/or the coating might occur, especially if the light has a significant near-ultraviolet and blue component. Photocopying of these objects should be restricted. The need for more research on this topic is evident.*

Fluorescent Whitening Agents

Both inorganic and organic compounds can act as fluorescent whitening agents (also known as fluorescent brightening agents or optical whiteners or brighteners). In order to function successfully as an optical brightener, a substance must absorb (near-)ultraviolet light and emit blue fluorescence with a good quantum yield. The most common inorganic fluorescent brightener is titanium dioxide; zinc oxide has also been used extensively. Optical brightening by organic compounds has been reviewed by, among others, Evans (1980), Leaver and Milligan (1984), and Moura, Oliveira-Campos, and Griffiths (1997). A large portion of the primary literature is in Japanese or in patent publications, so accessibility of experimental details is often very limited.

Inorganic compounds

The use of titanium dioxide as a fluorescent whitening agent is very common, but the fact that it can also function as an efficient photocatalyst in a wide range of situations (see, for example, Kamat 1993) may provide some problematic situations for collections care professionals in museums. When titanium dioxide is present as a finely divided powder that acts as an optical brightener in papers, in textiles, or in the supporting layer for photographic emulsions, it can also interact with media applied to the supports or with the support materials themselves. Titanium dioxide and other inorganic fluorescent whitening agents can serve as photocatalysts for oxidative degradation reactions of colorants. For example, titanium dioxide has even been selected to demonstrate photosensitized bleaching of dyes in a general chemistry demonstration (Giglio, Green, and Hutchinson 1995). Titanium dioxide or zinc oxide might also interact with inorganic ions present as impurities in the support or colorants, further enhancing the catalytic effect. The research reports discussed here are relevant to these possibilities.

Wilkinson and Kelly (1989) examined the effect, as measured by diffuse reflectance flash spectroscopy, of irradiating zinc oxide powder doped with copper, nickel, cobalt, iron, or manganese using a 532 nm laser pulse of less than 200 mJ of energy. Transient absorption spectra of tens of microseconds' duration were observed in the visible region. They were indicative of oxidation–reduction reactions involving the dopants. Each laser pulse used in these studies delivered about 5×10^{17} photons in a few tens of nanoseconds. This can be compared to a xenon strobe, which would irradiate an object with roughly the same number of photons in the 530–540 nm range with each flash that lasts 0.01 second. Because the irradiation of the object by the photoflash is spread out over a much longer time than that of the laser pulse, the concentration of excited metal atoms at any instant is less than 0.00001 times as great as that generated by the laser pulse. The quantum yield for photochemical reactions in pigment mixtures containing iron and zinc oxide is unlikely to be great enough that light from xenon flash photography would cause significant change. Paper that has iron inclusions as an impurity and also contains zinc oxide as an optical brightener is similarly unlikely to be affected by photoflash.

In a study of the kinetics of fading of opaque alizarin lake films containing titanium dioxide, Johnston-Feller and coworkers (1984) found that the fading process was essentially independent of the ratio of titanium dioxide to pigment at various pigment concentrations. The light source was a xenon arc in an Atlas Fade-Ometer, with ultraviolet light above 320 nm included. The smallest amount of titanium dioxide used was 2.5% pigment volume concentration. A yellow intermediate in the fading of the colorant was detected only if the fluorescent brightening agent was present in the lake formulation. The presence of this intermediate caused a color shift during the early stages of fading. The rate constant for disappearance of the yellow compound is greater than that for overall fading of alizarin. The report does not provide any additional data beyond what was presented earlier for these materials.

San Román and his colleagues (San Román 1996; Hodak et al. 1996) have reported on the use of red-light-absorbing phthalocyanines adsorbed onto titanium dioxide as photosensitizing systems. They have shown, for example, that in the presence of these dyes immobilized in this manner, red light can be used to cause photochemical reactions of compounds such as quinones that normally would not be sensitive to visible light of this low energy. The results for their experimental systems suggest that about 10^4 flashes of a studio xenon strobe with all ultraviolet wavelengths blocked would be able to cause an easily perceptible change. Although the experimental systems are not exact equivalents of actual objects found in museum collections, the results are suggestive of effects that may possibly arise when metal phthalocyanine dyes and quinoid dyes are used in combination on papers or fabrics containing titanium dioxide as an optical brightener.

The photolysis of several xanthine, oxazine, and thionine dyes adsorbed onto titanium dioxide powders or colloids has been studied by diffuse reflectance techniques (Fox and Dulay 1996). The researchers observed that, in general, the rates of secondary dark reactions of the dyes were greater on the powders than in the colloids. Most of the experimental results were obtained using near-ultraviolet light of wavelengths less than or equal to 370 nm. In the case in which data are presented for changes induced by 532 nm irradiation, the pH of the sample suspensions was 4.0 or less, and there was a huge excess of photons over dye molecules. Although the qualitative observations made in this investigation may be relevant to dyes on museum materials that contain titanium dioxide, specific calculations on the effects of photoflash cannot be made.

Tennakone and coworkers (1996) report that both purified and commercial tannins will complex with nanocrystalline, microporous titanium dioxide and participate in photoelectrochemical processes. The action spectra for observed photocurrents were found to be similar to the absorption spectra of the dyes. The authors also remark that tannins form strong complexes with Fe(III) and that these complexes are not adsorbed onto titanium dioxide and, therefore, are not active. Numerical data are inadequate to perform calculations based on the information in this report. The qualitative results may be relevant to possible photocatalytic effects, however, if inks containing tannins (but not iron compounds) are present on paper brightened with titanium dioxide.

A group led by Liu (Liu, Hug, and Kamat 1995; Nasr et al. 1996) has investigated photoinduced electron transfer processes that occur when organic dyes, such as cresyl violet and rhodamine 6G, are adsorbed onto aqueous colloids of the semiconducting materials silicon dioxide, SiO_2, and stannic oxide, SnO_2. In the case of cresyl violet, the dye was chosen because of its use as a sensitizer in color photography. The researchers excited the mixtures with laser pulses at 355 nm (for cresyl violet and sensitizers) or 532 nm (for rhodamine 6G). In the latter case, they were able to detect photocurrents generated with about 1% efficiency. This is the same order of magnitude as the quantum yield for photochemistry of the dye in solution, and much higher than the yield in the solid state. The results suggest that if these dyes are adsorbed onto semiconductor-type materials in a museum object, the possibility for photochemical reaction may be much greater than is usually expected.

> **Conservators' Note:** *The role of the inorganic fluorescent whitening agent titanium dioxide in promoting photodegradation of media and supports in museum objects cannot be evaluated quantitatively on the basis of the information in the literature, as exemplified by the reports reviewed above. The data are suggestive of possible deleterious effects that could impact exposure to photoflash as well as total display lifetimes of objects in museum and archival collections. There is a need for research on this topic, with experiments designed from the point of view of the conservator.*

Organic compounds

Organic substances that function as optical brighteners contain conjugated double bonds, but not such an extensive conjugated system that they absorb visible light and appear colored. Examples of classes of organic compounds used as fluorescent whitening agents are pyrazolines and stilbene derivatives. Although these substances have a high fluorescence quantum yield, they are also subject to eventual photodegradation; that is, they also have a low quantum yield for photochemical reactions. Some research has been conducted on the photostability of organic optical brighteners themselves. These investigations tend to be conducted with ultraviolet light sources, so the results are often not useful for performing calculations of photoflash exposures. Examples of some relevant research in this field are mentioned here.

Shosenji, Yamada, and colleagues have investigated the effects of adding pyrazoline derivatives to aminoanthraquinone and triphenylmethane dyes in solution, on acrylic polymers and copolymers, and in nylon and cellulose acetate films (Yamada, Shosenji, and Gotoh 1977; Yamada et al. 1981; Shosenji et al. 1983). The majority of these studies were performed with 360 nm near-ultraviolet light, which was absorbed efficiently by the optical brightener but not by the dye. In most cases, the presence of the dye increased the rate of photodegradation of the fluorescent brightener. More important, the presence of the optical brightener markedly increased the fading of the dye by the near-ultraviolet light. The limited numerical data provided permits an approximate calculation suggesting that about 10^6

flashes could be safely tolerated by the indicator dyes used with the pyrazoline fluorescent brightening agents in an acrylic copolymer object at a concentration of about 10^{-5} molar (see chap. 3 for more details).

Evans, Rivett, and Waters (1976) studied the effect of pyrazoline fluorescent whitening agents on the rate of yellowing of wool. This rate increased with optical whitener concentration up to a concentration of 0.4% and then leveled off. The effect is easier to monitor experimentally if the yellowness index of the wool is measured without exciting the fluorescence of the pyrazolines. This study was done using light from a mercury arc that included all wavelengths greater than 295 nm, so the results cannot be used to calculate the effects of photoflash.

Davidson and coworkers (1985) compared the photochemical stabilities of two pyrazolines to that of commercial whitening agents on wool. They observed that the stability of the brightener in solution was not a good predictor of its relative fastness on wool. This was at least partly due to significant yellowing on the surface of the treated fibers. The discolored layer then partially screened the underlying brightener molecules from the incident light. The results observed did not permit the authors to determine whether wool degradation products were involved in the breakdown of the whitening agents. Near-ultraviolet light from a black-light fluorescent lamp was used in this study. The maximum emission wavelength of 350 nm for this lamp precludes extending the results numerically to the effects of photoflash.

Leclerc and Flieder (1992) compared the changes caused by artificial aging of several commercial papers and cotton blotting paper, each with and without a commercial optical brightener used by the manufacturer. These brighteners were not characterized chemically. The samples were aged by xenon-arc exposure for 150 hours on each side of the paper. The total dose was 18 J/cm^2, which was sufficient to fade Blue Wool no. 5. This relatively large irradiation caused much more yellowing (measured by CIE brightness values) of the samples that contained the brighteners than the control papers without them. Because the controls were so much less bright initially, however, the samples containing the fluorescent compounds still appeared brighter than their respective controls after the light-aging process. The authors suggest that some of the observed overall effect may have been due to degradation of the optical brighteners themselves. Changes in mechanical properties of the blotting paper without the brightener and in the degree of polymerization of the cellulose molecules in the commercial papers were minor. However, the unsized blotting paper sample that contained the fluorescent brightener was significantly chemically degraded by the light-aging process. Because the light exposure was so great and information on the spectral output of the light source is not available, calculations relevant to photoflash exposure cannot be carried out.

Leaver (1978) investigated the photodegradation of wool in the presence of stilbene fluorescent brighteners, irradiating the fiber samples with a fluorescent black light. When the wool was wet, about 25% of the amino acid tryptophan was degraded by irradiation with roughly 10^{19} $photons/cm^2$. The degradation was significantly less when the fiber was dry, which is consistent with an earlier report by Milligan and Holt (1974) that stilbene fluorescent brighteners were ten times more stable on dry wool than on wet. The particular derivatives these latter authors studied were more than ten times as stable on wet wool as in solution. Because the light sources in these studies emitted radiation mainly in the band 350 ± 30 nm, these results cannot be used numerically for the present investigation. In a citation of earlier work from the same laboratory, the wavelength for maximum yellowing of wool containing a stilbene fluorescent brightener was identified at about 390 ± 50 nm. Also, about 20% as much yellowing was observed when the samples were irradiated with a band of visible light at 520 ± 15 nm. Because light doses are not given in this article, calculations of flash exposures cannot be made.

Oda, Kuramoto, and Kitao (1981) studied the photofading of stilbene fluorescent whitening agents sensitized by rose bengal. When they used light of wavelengths greater than 510 nm, which was absorbed by the sensitizing dye only, they observed about 15% fading of the stilbene in 10 minutes. Information provided on the light source can be used to calculate that no more than 10^4 flashes would be the maximum allowable for an object containing a stilbene optical brightener and a dye that could sensitize the degradation of this brightening agent (see chap. 3 for more details).

Davidson, Ismail, and Lewis (1987) investigated the breakdown of wool by sulfonate derivatives of stilbene optical brighteners by testing model systems as well as wool samples in their studies. Results for indole derivatives (model compounds for the amino acid tryptophan, which is photodegraded in wool in the presence of optical brighteners) in solution and for the stilbene derivatives and a commercial whitening agent indicated similar extents of sensitized photodegradation of wool. The numerical results cannot be extended to photoflash effects because the light source used emitted mostly ultraviolet wavelengths below 360 nm.

Yagami and coworkers (1988) have compared the fading of azo dyes and stilbene fluorescent brightening agents on wool or on cotton irradiated with light sources emitting different relative amounts of ultraviolet radiation, including mercury and xenon arcs with Pyrex filters. In general, the dyes partially protected the fluorescent brightening agents from degradation; and for some dyes—C.I. Direct Blue 15 on cotton, for example—the optical brightener partially protected the dye. A nickel salt that scavenges singlet oxygen increased the loss of whitening agent on cotton and decreased it on wool. Again, because all the light sources contained a significant component of near-ultraviolet light below 360 nm, the results cannot be used to predict the effect of photoflash exposures.

Reagan and coworkers (1989) studied the effects of stilbenes and other organic fluorescent brightening agents on the photoyellowing of silk, using exposure to a xenon arc in an Atlas Weather-Ometer to mimic the yellowing effect of daylight through glass. They found that although one stilbene compound and a coumarin derivative slightly accelerated the rate of photoyellowing, none of the five chemicals they tested had an effect over long periods. No data relevant to the possible effects of flash-lamp exposures were presented. A group of researchers including the same authors (American Association of Textile Chemists and Colorists 1990) also reported that silk samples dyed with a variety of triphenylmethane or azo dyes and treated with stilbene, coumarin, or triazole fluorescent brightening agents displayed fading behavior dependent on the dye rather than on the brightener. Thus, the triphenylmethane dyes, which were less lightfast than the other colorants, were also less lightfast in the presence of the optical brighteners.

> **Conservators' Note:** *In general, research on fluorescent whitening agents has employed light sources with much greater near-ultraviolet output and shorter ultraviolet wavelengths than those emitted by xenon photoflash lamps. This has been done mainly to study the role of the light energy, absorbed by the brightener itself, that subsequently leads to fluorescence. The results of these investigations are thus not particularly useful with regard to effects of photoflash. However, they may be qualitatively relevant to the effects of photocopier sources on modern commercial papers, which, almost without exception, contain optical brighteners. Additional research specifically designed to determine the consequences of photocopying and of extensive exposure to photoflash on materials containing a variety of optical brighteners in the presence and absence of colorants should be performed.*

Photographic and Reprographic Materials

Photographs present a tremendous challenge for light-exposure decisions because of the great variability among individual objects created ostensibly from the same materials and by identical procedures. The photochemical and chemical processes involved during the exposure and development of photographic emulsions are highly complex, providing many opportunities for variability. See, for example, section 8.6 in Wayne (1988) and several chapters in *Neblette's Eighth Edition* (Sturge, Walworth, and Shepp 1989). In the process of creating unique objects, artists may have further complicated the situation by purposely varying procedures from those recommended by manufacturers. The quantum yield for change in silver halide crystals in photographic emulsions is on the order of unity. Furthermore, relatively low concentrations of metallic silver in association with silver halide crystals that inadvertently or intentionally remain in the emulsion after exposure and development cause an increased rate of silver deposition with additional light exposure. These facts increase the likelihood that an improperly developed photograph will not be stable to light.

Some types of photographs, for example, black-and-white gelatin silver prints on high-quality paper, have been categorized as generally durable (Colby 1992). These prints are expected to have a lightfastness rating similar to Blue Wool no. 7 when they are glazed with ultraviolet-blocking material. They should, hypothetically, not be adversely affected by extensive exposure to photoflash, provided that all ultraviolet light is removed from the light source. As the earlier comments and the evaluations that follow indicate, however, this standard cannot be applied indiscriminately.

In contrast to many gelatin silver prints, other types of photographic objects (for example, nineteenth-century halide-fixed images and some color prints from the 1970s) have proven to be so light sensitive—sometimes unexpectedly so—that any proposed light exposure should be carefully considered and monitored. An additional complication to caring for photograph collections is that photographic materials may change in the dark, especially if low-temperature storage is not used, and that this behavior may or may not be affected by prior light exposure.

> **Conservators' Note:** *In the absence of evidence to the contrary for particular items, caretakers responsible for photographic collections would be safe to assume that photographs are light sensitive. For all photographic materials, the scrupulous removal of ultraviolet wavelengths from light sources is essential.*

Historic photographs

Halide-fixed images

Some types of very early photographic materials, such as W. H. F. Talbot's halide-fixed images (Moor and Moor 1992; Ware 1994), are among the most light-sensitive objects found in museum and archival collections. Awareness of the extremely fugitive nature of these objects is widespread, and conservators generally attempt to minimize the exposure of these works to light in all circumstances. Precautions should always include forbidding gallery flash photography and limiting studio photography. Consideration should be given to creating a reproduction for exhibition; a stand-in can be used for the setup so that the original is subjected to the minimum required number of flashes with complete ultraviolet filtering. Ware calculated that the threshold of detectable change would be caused by about 4×10^3 flashes of light at an intensity appropriate to a film speed of 100 ASA units and an aperture of *f*-8. Using the assumption that flash exposure should be limited to 0.01 of what causes a noticeable change, forty such flashes from a studio strobe could be allowed over the lifetime of the object. The same order of magnitude of flash exposure can be calculated from data presented by Moor and Moor.

> **Conservators' Note:** *Ware cautions against monitoring the appearance of highly sensitive photographic objects by densitometry or colorimetry because each measurement requires exposure to a rather powerful light source for about one second. He recommends, instead, periodic comparison with a many-step (presumably stable) gray scale specially prepared for each object.*

Lippmann photographs

The relative stability of historic images made by the Lippmann color process,[2] in which the emulsion is deposited as distinct layers and does not contain colorants, stands in contrast to the light sensitivity of halide-fixed images and some other early photographs. The image in a Lippmann photograph is generated by the interference of particular wavelengths of light that are reflected by the individual layers in the emulsion. This is the same principle as that on which holograms are based (Fournier and Burnett 1994). The colors result when some visible light waves reinforce each other and others cancel out. The preservation of Lippmann photographs depends on maintaining the mechanical stability of the polymeric emulsion layers (which were usually gelatin) and of the fine grains of colloidal silver embedded in them.

> **Conservators' Note:** *Moderate photoflash exposure should not adversely affect the essential properties of Lippmann photographs, provided heating is avoided.*

Albumen photographs

Pretzel (1992) monitored appearance changes in a collection of nineteenth-century albumen prints, some of which were extensively displayed. Light exposures ranged from 0 to approximately 2.4×10^5 lux hours. Because the only information given about the light sources is that intensity was kept below 50 lux at the different venues, specific calculations with respect to photoflash exposures cannot be made. Although there was significant variability in the response of areas of different densities in the same and different photographs, some general trends were detected in the data. The prints tended to darken in the dark, with the extent of change being greater with greater initial density. After one and a half years, the densest areas had darkened by a noticeable amount, so that the contrast of the prints had increased. The overall change in contrast of some of the stored photographs was as great as or greater than the change in the displayed prints. In other instances, the exhibited prints bleached during the first period of display and then began to darken. The overall change after approximately 9×10^4 lux hours of light exposure was barely noticeable. This phenomenon may be due to different responses of the support, protein, and image to light over time. Unfortunately, data that would permit further evaluation of this behavior are not given. The significant overall changes in contrast that were observed for many of the prints (in the categories of full display, partial display, and dark storage) underscore the necessity of evaluating proposed light exposures on a case-by-case basis.

> **Conservators' Note:** *The fact that detectable change had occurred after 2.2×10^4 lux hours in one of the prints on display suggests that exposure to flash might best be limited for historic albumen photographs. On the basis of the information available, it is not possible*

to predict which albumen prints might show significant deleterious changes.

Cyanotypes and blueprints

When made by the wet method, cyanotypes and blueprints are recognized as being able to recover from bleaching by light if subsequently kept for a period in the dark (e.g., Fireman 1997). The color changes are due to shifting oxidation–reduction equilibria among iron and potassium salts of the anions ferricyanide and ferrocyanide (Prussian blue is mostly iron ferrocyanide) (Kosar 1965; Dessauer and Looney 1989; Berrie 1997). Long-term effects of light exposure on the support (which is not necessarily material with good light stability) may interfere with the process or change the appearance of the print. Furthermore, as Fireman points out, the limits of recyclability of these objects have not been investigated. The ability of aged blueprints or cyanotypes to recover after they have been exposed extensively to light, or after they have undergone a large (and usually unrecorded) number of photochromic cycles, has not been investigated experimentally. Additionally, historic Prussian blue is known to be less stable than the modern pigment (Kirby 1993). There is no quantitative information in the literature that would indicate whether exposure to photoflash would interfere with the ability of these historic prints to recover their color in the dark after photobleaching. Another process for making blueprints, the semidry method, leaves some unreacted salts in the paper, making such blueprints more apt to fade permanently in light. Experimental research on the response of blueprints to light would provide useful insight.

Other historic photographic prints

Another historic photographic procedure for making color prints is the gum bichromate process. The emulsion consists of gum arabic containing sensitizer and a watercolor pigment. A contact print is made by placing the object on the paper containing the emulsion and then exposing it to light; the print is then developed in plain water (Scopick 1978). The light hardens the exposed emulsion, which remains after washing. The sensitizer is a dichromate salt.

> **Conservators' Note:** *The lightfastness of gum bichromate prints will generally be determined by the watercolor pigment used in the emulsion. If it is a relatively stable organic compound or an inorganic pigment, the prints should not be adversely affected by moderate exposure to photoflash.*

McElhone (1992) has briefly reviewed the effect of visible and ultraviolet light on the various components of historic and modern photographs and has monitored with densitometry the effects of display lighting conditions on a variety of photographic prints. A salted paper print, several different types of albumen prints, and two different gelatin silver prints (printed-out print [POP] and wax-coated) showed no significant change when subjected to 3×10^4 lux hours of tungsten light.

This result suggests that these particular photographs could be exposed without significant effect to more than 10^3 flashes from a xenon strobe lamp with the ultraviolet blocked. McElhone also observed that several gelatin silver prints from various studios, including one without toning, and some vintage albumen prints did not change significantly after 8×10^5 lux hours or more of exposure. This observation suggests that these prints might be exposed to on the order of 10^5 flashes without incurring detectable damage. Other examples were much more light sensitive, however. A modern dye-coupler print was faded significantly after 3×10^4 lux hours of exposure, and a modern gelatin silver print and a gold-toned albumen print each showed significant density increases after about 2×10^5 lux hours. The latter samples continued to change over time in the dark. These data suggest that several thousand photoflashes from a xenon lamp would cause a significant change, so flash exposure of these sensitive objects might best be limited to about 100 flashes.

> **Conservators' Note:** *An important lesson from the results reported here is that the type of photograph does not determine its stability to light; ideally, photographs should be considered on an individual basis.*

Modern photographs

Black-and-white prints

One group of modern black-and-white photographs has tended to show unique problems related to their composition: resin-coated prints from the 1970s in which the emulsion layer and the backing are coated with a film of polyethylene. These prints frequently developed colored spots of varying size and distribution in the polyethylene layer. It has been suggested that the presence of titanium dioxide has a bearing on this behavior (see Wilhelm 1993, chap. 17). Yellowing and the appearance and development of the spots do not seem to follow the reciprocity rule. They may not appear for a long time, on the order of several years, and then develop suddenly. Wilhelm conducted experimental investigations of these effects over a period of years, mostly with fluorescent light sources lacking ultraviolet blocking. The changes occur more slowly, but are not prevented, if ultraviolet blocking (specifically UF-3 glazing) is used. A reasonable order-of-magnitude flash exposure cannot be calculated for 1970s resin-coated black-and-white prints on the basis of the data available from this study.

Color prints, slides, and transparencies

The chemical processes involved in color photography were recently reviewed, with excellent illustrations, by Theys and Sosnovsky (1997). Zollinger (1987b) briefly discussed the chemistry of colorants used in photographic and some reprographic systems of the 1980s. Keller (1993) notes that the dyes used in color photography are much less stable to light than their supports; he also comments that the desired lifetimes of color photographs are very much longer than the lifetime of a particular technology used. With these caveats in mind, the conservator dealing with color photographs needs to attempt to identify the approximate date of a photograph and the process used to produce it before specific decisions can be made regarding a useful display lifetime and the feasibility of unrestricted flash photography. The differences in the data and the conclusions reached in the following discussions illustrate the problems associated with color prints, slides, and transparencies.

Giles and coworkers (1973) compared the fastness of five commercial color prints and of color-coupled transparencies made with several different speeds of film to the Blue Wool standards, using a Microscal lamp, projector lamps, or indoor daylight. These light sources radiate different amounts of near-ultraviolet light. The researchers found that the sensitivity of the transparencies to light increased as the film speed increased, regardless of manufacturer, whereas the lightfastness of the prints varied only slightly with the source of the print papers. Transparencies made with very fast film and instant-developed color paper prints had a light sensitivity equivalent to Blue Wool no. 1. Three of the five color prints and the transparencies made with fast film had sensitivities between nos. 1 and 2. Five hours of exposure to the Microscal lamp caused a just-perceptible change in the samples ranked no. 1. Using these data to determine photoflash exposure indicates that the most sensitive transparencies could be exposed to about 10^4 flashes from a xenon strobe lamp without the ultraviolet blocked over their entire display lifetime (see corresponding section of chap. 3 for calculation details).

> **Conservators' Note:** *Protection of prints and transparencies with a glazing that blocks ultraviolet light and use of flash sources with ultraviolet wavelengths scrupulously blocked would suggest that more than 10^4 flashes could be allowed for 1970s color materials. The exact amount cannot be calculated, however, because the specific effect of near-ultraviolet wavelengths on the experimental samples was not identified.*

The light stability of developed color patches in a variety of commercial color reversal films exposed to a projector lamp was studied by Fujii, Fujii, and Hisanaga (1988). The most sensitive samples tested were the magenta and cyan patches on a Kodachrome film, which showed a ΔE of about 5 after a 20-minute exposure.

> **Conservators' Note:** *Based on the output of the lamp, this observation suggests that the most sensitive film could withstand, without deleterious change, at least a few thousand flashes from a xenon strobe without near-ultraviolet filtering. Because the spectral power output of the quartz–halide projector lamp is not known in the ultraviolet, however, this number could be a significant underestimate.*

Townsend and Tennent (1993) reexamined the lightfastness of contemporary color transparencies from four manufacturers. They made the slides by photographing the gray side of a Kodak neutral test card. They then exposed the slides in a Microscal unit with either the emulsion side or the base side facing the light source, with and without ultraviolet filtering. Their purpose was to determine the storage and display lifetimes of these materials when they are used for documentation purposes or in before-and-after-illumination didactics, so they subjected the sample slides to large total exposures. When ultraviolet filtering was in place, the worst-performing transparency changed by a ΔE of about 25 units, whereas ΔE of 1 corresponded to a noticeable change. The authors supplied enough experimental details to allow calculations of flash exposures to be made. The most sensitive slides could be subjected to about 5×10^3 to 1×10^4 flashes from a xenon strobe lamp with ultraviolet filtering (see the corresponding section of chap. 3 for more details).

> **Conservators' Note:** *When near-ultraviolet light was present in the light source for the study, some of the sample slides showed a* ΔE *that was ten times greater than in the absence of ultraviolet light. This suggests that exposure to photoflash sources without near-ultraviolet filtering should be limited to fewer than* 10^3 *flashes over the lifetime of the transparencies.*

Wilhelm (1993, chap. 6) has reported severe reciprocity failure for color slides, especially for the magenta dyes. Most slides faded more when subjected to low-intensity, long-term exposures than to high-intensity, short-term exposures. In a few cases, the reverse was true, suggesting that once the transparency has been exposed to light, it continues to fade in the dark. It is not clear if final measurements on all samples were made after the same total elapsed time. The experimental exposures used in this investigation were performed with the slides 3.05 m (10 ft.) from a tungsten–halide projector lamp without any filtering. This light source has a significant component of ultraviolet light below 365 nm, so the results cannot be easily extended to the effects of photoflash. The nonreciprocity findings indicate that further investigation would be appropriate.

Frey and Gschwind (1994) measured the spectral changes in cyan, magenta, and yellow dyes in various color transparencies after exposure to 2.5×10^4 lux from fluorescent lights of unspecified type for at least 400 hours. This total of 10^6 lux hours of exposure caused a 50% loss of magenta dye in the lightest density area tested in one case. Assuming (1) that the fluorescent light was of the daylight type; (2) that a 5% loss of the dye would be the smallest that is noticeable and thus represents the display lifetime; and (3) that photoflash exposure should contribute less than 0.01 of this, it can be calculated that about 10^4 flashes of a xenon studio strobe lamp without near-ultraviolet filtering might be the maximum allowable. The changes these authors observed in other dyes and in print media were all much smaller.

Fading of cyan, magenta, and yellow patches on five brands of commercial color photographic paper was investigated by Siripant (1986), who exposed samples in a weathering apparatus for at least 16 hours. The illuminance in the visible at the surface of the papers was just under 2×10^5 lux, so the smallest exposure was again over 3×10^6 lux hours of visible light, plus about 3×10^{20} near-ultraviolet photons/cm^2. These relatively large exposures caused just-perceptible changes in some of the magenta and cyan patches and would therefore result in color differences in actual prints.

> **Conservators' Note:** *Using the assumption that photoflash should contribute less than 0.01 of a perceptible change, the numbers suggest that about 10^4 flashes from a xenon studio strobe without near-ultraviolet blocking might be the allowable lifetime flash exposure for 1980s color prints. Data published by Wilhelm in an earlier report (1981) in which he monitored red, green, and blue density changes in a resin-coated color print exposed to fluorescent lights lead to a similar conclusion. Wilhelm stressed the importance of monitoring areas of low density, because a given amount of degradation of dye molecules in these areas will lead to a more noticeable change.*

Wilhelm (1993, chaps. 2, 3) presented more data similar to that above and much additional evidence for the instability, over the long term, of color prints exposed to fluorescent lighting. In general, the magenta dyes were the most unstable, but they were less likely to show reciprocity failure. Some cyan dyes faded more from exposure to the incandescent light than to the fluorescent lights used in Wilhelm's study. A lot of evidence is presented that supports the nonreciprocity of fading of some materials. The 1970s resin-coated paper prints have a tendency to yellow, which contributes to the nonreciprocity observed. Because Wilhelm mainly addressed long-term stability and used long exposures to light sources without near-ultraviolet blocking, much of the data is not useful for calculating photoflash exposures. In the poorer-quality photo papers tested, the rate of fade of the magenta dye appeared to increase after the first year of exposure. This finding contradicts the observation mentioned earlier that the magenta dyes are less likely to show reciprocity failure. The author states that the results were similar whether or not a UF-3 ultraviolet-blocking filter was used.

In the same publication, Wilhelm (1993, chap. 4) summarizes the effect of supposedly protective coatings on color photographic prints. He determined that the coatings do not provide significant protection from light over the long term. They are likely to become yellowed themselves, thus contributing to the apparent change in appearance of the print. In some instances, the coatings may actually induce dye fading. Wilhelm used large exposures to collect the data that led him to these conclusions. The results do not affect the approximations presented above regarding photoflash exposures.

> **Conservators' Note:** *The behavior of these color prints suggests that visible light contributed significantly to their*

> *fading, so a calculation of photoflash exposure is appropriate. Assuming again that 0.01 of a just-perceptible change is the maximum that should be contributed by flashes, about 10^5 flashes from a xenon source with the ultraviolet blocked might be permitted over the display lifetime of the type of color prints investigated by Wilhelm. It should be kept in mind that exposure to light may in some cases lead to further alterations after a print is stored in the dark. No useful numerical data are given on this subject, which might be an appropriate topic for further study.*

Bergthaller (1985) has reported relatively high light stability for several metallizable azo dye derivatives used in Cibachrome prints; he based this on a comparison with Blue Wool standards. The print materials were exposed in a Xenotest apparatus, but no experimental details are provided. Stabilities equivalent to Blue Wool nos. 5 and 6 are reported, suggesting that Cibachrome prints may be unaffected by exposures of up to about 10^6 flashes from a xenon camera flashlamp.

McElhone (1992) has called attention to the fact that several researchers have reported different amounts of fading (or times to a perceptible fade) for apparently similar resin-coated prints and for dye-coupler color prints from different laboratories when they were exposed to various tungsten light sources. These observations could be due to variations in the spectral distribution of the light emitted by the light sources, differences in the conditions and the materials (dyes, emulsions, and supports) of the prints, or both. Further research would be needed to clarify these observations and to determine how the physical state of the various noncolored materials might affect the appearance of the colors in the photographs.

Dye photochemistry

Concern for the light stability of the dyes used in color photography has spurred studies of their photochemistry. Douglas and Townsend (1992) have reviewed the principles of photophysics and photochemistry as specifically applied to azomethine dyes used to form color images in photographic films and prints. The authors state that the quantum yield for photodegradation of these compounds is usually about 10^{-6} or lower.

Exceptions to this conclusion have been observed, however. For example, Chatterjee and coworkers (1990), while investigating the solution photochemistry of carbocyanine phenyl borate derivatives, observed quantum yields for photochemical alterations as high as 0.1 for one compound when it was irradiated with visible light at 532 nm. The data permit an approximation that a maximum of about 10^3 flashes could be allowed (see corresponding section in chap. 3 for more details). Chibisov and colleagues (1995) observed bleaching in 10 seconds during a flash photolysis study of some polymethine carbocyanine dyes in nonpolar solvents, but they do not provide enough experimental details for calculations.

Another example is provided by the results of Wilkinson and coworkers (1993), who investigated the photochemistry of magenta pyrazolotriazole azomethine dyes in gelatin coatings. Dyes of this type are present in color photographic emulsions. One of several similar dye derivatives was observed to degrade sufficiently that the sample had to be replaced after every 100 laser flashes having less than 10^{-5} J of energy per 585 nm flash. Calculations based on the sample preparation data suggest that the yield for photodegradation at this wavelength may be as high as 0.02 (see chap. 3 for more details).

> **Conservators' Note:** *The maximum allowable number of photoflashes from xenon studio strobes with ultraviolet blocking would be on the order of 10^3 for prints containing this magenta dye. It should be noted that other similar dye derivatives used in this study did not display the same degree of photosensitivity.*

There is a substantial literature on dyes of these types published by employees of companies that make and market photographic materials (e.g., Smith, Herkstroeter, and Eddy 1976; Giacherio and Valente 1985; Furuya et al. 1994). It is difficult to relate the information in these publications to the current topic because for proprietary reasons the authors usually omit numerical experimental details or quote all data in relative terms without providing a standard for comparison, or both. Giacherio and Valente did mention that the yellow staining sometimes observed in the magenta coating layer may in some cases be due to dark reactions of a dye photooxidation product in the layer.

Reprography and reprographic materials

The technical aspects of several modern reprographic processes that produce paper-based copies (photolithography, laser and ink-jet printing) have been described by Gregory (1994). Reports on various aspects of new and developing technologies may frequently be found in conference proceedings of the Society of Photographic and Imaging Engineers and the Society for Imaging Science and Technology. Jones (1990) has summarized materials and processes involved in Xerox, Thermofax, and other similar reprographic systems. He mentions that ultraviolet radiation can catalyze the oxidation of toner components. Theys and Sosnovsky (1997) briefly summarize some contemporary color reproduction processes. Black-and-white and color xerography and laser printing processes are succinctly described in a brochure published by the Australian Archives (Photocopying 1993); the brochure also summarizes permanence properties of the materials generated by these processes. Jarry (1996) and Orlenko and Stewart (1998) have addressed the identification of various types of contemporary prints generated by computer-controlled printing processes, based on the appearance of surface features under low magnification. Wilhelm (1993, chap. 1) commented on the poor light stability of Iris printing inks used in ink-jet printers and mentioned that more-durable versions were expected to be available within a year of publication of his monograph. Anecdotal evidence of the relative instability of some colorants in Iris prints abounds, but published papers with numerical data could not be located.

Dye systems

Photochemical studies of some of the dye systems used specifically in reprographic processes can be found in the chemical literature, but they are frequently concerned with the development of the reproduction and not its subsequent photostability (see, for example, Zollinger 1987b). Investigations addressing photostability have often employed only near-ultraviolet radiation for the artificial aging of experimental samples, so the results are not relevant to the present inquiry. A few examples with some of the more useful data are discussed here.

Kuramoto and Kitao (1982) and Allen, Hughes, and Mahon (1987) investigated the fading of the crystal violet lactone color-forming system that has been used in carbonless copy paper and in gravure printing. Kuramoto and Kitao, observing the behavior of the dyes deposited on silica gel, found them to be about ten times as stable as crystal violet in solution (see under "Triphenylmethane dyes," earlier in this chapter), and to be partially protected by singlet oxygen quenchers that are transition-metal complexes. Allen, Hughes, and Mahon irradiated their samples with light from a xenon flashlamp (a Microscal unit) or a Xenotest instrument in the absence or presence of various agents intended to protect the dyes from photodegradation. In contrast to Kuramoto and Kitao's results, they observed protection by peroxide scavengers but not by singlet oxygen scavengers and noted that some stabilizers actually promoted fading. (Details of this 1987 study of the color-forming system based on crystal violet are presented in the corresponding section in chap. 3.)

> **Conservators' Note:** *On the basis of data supplied in the Allen, Hughes, and Mahon report, it can be calculated that about 10^4 flashes from a xenon studio strobe setup without ultraviolet blocking would be the maximum allowable over the lifetime of a document made with the crystal violet lactone color-forming system. Photocopying would be more deleterious and should probably be limited to a few hundred copies total.*

Oda and Kitao (1990) studied the effects of other types of stabilizers on the same color-forming system, preparing samples by deposition from solution onto microcrystalline cellulose or by casting poly(vinyl alcohol) films on uncoated paper. They found that some zinc salts of organic acids could protect the color from major fading. However, they exposed their samples to all light of wavelengths greater than 300 nm from a high-pressure mercury arc for 3 hours, so their numerical results are not useful with respect to photoflash or photocopier exposures.

Pigments and printing inks

Schmelzer (1984) studied the effect of near-ultraviolet light on the photodegradation of some colored pigments used in printing inks. The rates of fading of C.I. Pigment Yellow 127 and C.I. Pigment Red 53:1 were observed to increase with time of exposure, whether or not ultraviolet wavelengths were removed from the light source. This means that the changes in appearance, as measured by ΔE, increased with increasing light exposure. On the basis of the data supplied in the report, it can be calculated that a 3-hour light exposure, causing a ΔE of 5, is the equivalent of about 10^4 flashes from a studio strobe with ultraviolet blocking. This change would easily be perceptible by eye.

> **Conservators' Note:** *Flash exposure of C.I. Pigment Yellow 127 and C.I. Pigment Red 53:1 should be kept to only about 100 flashes. It is interesting that the red colorant responded in the same way in Whitmore and Bailie's (1997) study of the fading of clear glazings. The result for the printing ink is in reasonable agreement with that for the glaze and suggests that in both formulations the colorant-plus-carrier mixture does not obey the reciprocity principle.*

Searle (1994) reported on the fading of two red and two yellow pigments used in commercial inks formulated for offset printing. The specimens, on coated paper of unspecified composition, were provided by the American Society for Testing and Materials (ASTM). The most sensitive ink in the study contained C.I. Pigment Red 53:1. This ink showed a ΔE of about 15 after 92 hours of radiation from a xenon arc with cutoff filters that removed light of wavelengths below 410 nm. The total allowable number of flashes over the lifetime of an object with this red printing ink can be calculated using the reflectance spectrum of the ink provided by Searle and assuming that (1) the xenon light source is equivalent to that in an Atlas Weather-Ometer, (2) a ΔE of 5 corresponds to a perceptible change for this opaque colorant, and (3) flash should contribute no more than 0.01 of this change. The calculation suggests that the total allowable number of flashes would be about 4×10^5 (see the corresponding section of chap. 3 for more details).

It is instructive to compare this result to the approximate number of photocopies that could be made without significant effect on the display lifetime of a document containing this red printing ink. Data in Searle's report indicate that the ink is sensitive to near-ultraviolet light above 320 nm. A color change in the experimental samples that was about four times as great as that observed for visible-light irradiation occurred in about half the time when ultraviolet light was included. Calculations based on this information and on the output of the xenon light source in a 1980s photocopier (see table 1.2) suggest that several hundred copies would be the maximum allowable over the lifetime of a document containing this red ink (see chap. 3).

> **Conservators' Note:** *Moderate flash photography using lamps with ultraviolet wavelengths completely blocked would probably not be deleterious to printing inks containing C.I. Pigment Red 53:1. These formulations may be as much as 10^3 times as sensitive to photocopier light*

sources (containing near-ultraviolet wavelengths above 320 nm as well as visible light). This example serves to illustrate how much more light can be delivered by photocopier sources than by filtered camera flashes. Also, the possibility of reciprocity failure for inks and coatings containing this colorant should be borne in mind.

Color photocopies, faxes

Fujii and Fujii (1992) made color test patches using several photocopier machines that employed various technologies; they then faded the samples in a xenon fading apparatus or with a mercury-arc lamp. They reported an unacceptable ΔE in the magenta patches made by two different Canon photocopiers, and in the yellow patch made by one Canon photocopier, after exposure to 3.0 J/cm^2 of light energy between 380 and 420 nm. This corresponds approximately to 6×10^{18} photons/cm^2. If a xenon strobe setup, which irradiates the object with the order of 10^{14} photons/cm^2 in this wavelength range, is to contribute less than 0.01 of the experimental exposure, the number of flashes would need to be kept below about 10^3.

Norville-Day (1994), prior to undertaking conservation treatment of some color photocopies and faxes, investigated the fading of sample materials by light from several Philips 94 fluorescent tubes with ultraviolet-blocking filters. The most fugitive color in the photocopies was a brown, and the black areas were more than twice as fugitive as the brown areas on the photocopy. Data supplied by the author permit calculation of allowable flash exposures (see chap. 3).

Conservators' Note: *Photocopies with the brown color could endure up to 1.5×10^5 flashes from a xenon strobe with ultraviolet light blocked; the black faxes should be limited to about 5×10^4 flashes over their display lifetimes at the ink coverages used in these samples.*

A report that appeared more than twenty years ago affords an unusual opportunity to consider the possible effects of a xenon photocopier lamp from the mid-1970s. Domergue and Asmus (1976) used photocopier lamps of this type to cause two layers of green oil-based paint over moleskin upholstery fabric to become easily removable from the original textile with a spatula. The authors report that delivery of approximately 4 J/cm^2 of visible and near-ultraviolet energy from the lamps caused major flaking and some charring of the paint layers. This does not take into account that the experimental use of the lamps may have included some heating of the object. The report does not mention any damage to the underlying fabric layer.

Conservators' Note: *Considering that the xenon photocopier light source from the late 1980s (see table 1.2) delivered 17.5 mJ/cm^2 per copy to the objects on the glass, 250 copies may expose an object to enough energy*

> *to cause flaking of old paint layers. These numbers again suggest that making more than a few hundred photocopies of objects with light-sensitive media should probably be avoided.*

Digital imaging

Digital imaging captures the image of an original either by a scanning process or by a photographic procedure (Besser and Trant 1995). Detailed descriptions of the techniques and processes used to accomplish such reproductions can be found in Kenney and Personius (1992), Aaland and Burger (1992), Williams (1994), and Besser and Trant (1995).

Vitale (1998) recently raised the question of the visible and ultraviolet light exposure caused by scanners. As for photocopiers, the length of time of the scan and the light source strongly affect exposure. Vitale's informal measurements of some scanners suggested that the total light dose may vary among instruments by a factor of about 20. In almost all cases, exposure from one or a few scans will be negligible. If multiple scans are anticipated, or if highly fugitive materials will be scanned, a more thorough consideration of the conditions would be warranted. More detailed characterization of scanner light exposures would be in order.

Several reports on specific digital imaging projects consider in depth the stability of the materials used to make the reproductions (see, for example, Bagnall 1995; Kenney and Chapman 1996). Surprisingly, these publications frequently do not consider the effect on the original object of the lighting used during the imaging process, although Bagnall does cite avoidance of ultraviolet light as one advantage of digital cameras over scanners for imaging. Ideally, total exposures and the spectral output of possible light sources should both be considered when an imaging method is selected. If the object being scanned is a book or library material of no intrinsic value—that is, if the sole reason for producing the image is the preservation of the information contained in the original—disregard for the possible effects of light exposure on the object may be justified. In other cases, the original object is itself of historic or artistic value (Bagnall, for example, worked on a project to create digital images of papyri), so this approach is not appropriate. It appears, however, that designers of imaging equipment and those planning imaging projects often show little concern for the effect of light exposure on the original documents being copied. A similar disregard for possible deleterious effects of light is evident in the proceedings of a 1986 conference on preservation photocopying (Preservation photocopying 1987). A proceedings paper by White (1987) described the extensive effort spent to ensure that the process of photocopying valuable books did not inflict mechanical damage on the documents, but the possible deleterious effects of light exposure were not considered during the conference.

> **Conservators' Note:** *The lack of concern for possible light-induced damage to original documents by imaging*

devices such as scanners does not mean that problems do not exist. Conservators should be aware of this situation, become involved in the choice of imaging procedure, and take steps to inform photographers and technicians of the light sensitivity of the materials being copied.

Messier (1997) recently posted on the Conservation Distribution List Web site an example of irradiation conditions during another kind of imaging process: laser digitization. This procedure might be undertaken, for example, to create a computer file on the three-dimensional surface features of an object. On the basis of the numbers Messier supplied, it can be calculated that approximately 10^{18} photons/cm^2 of red light at a wavelength of 633 nm would strike the surface of an object subjected to this procedure. This could exceed the amount of blue colorant molecules (which absorb red light) present by a few orders of magnitude. Thus, there is a possibility of causing unacceptable fading (see corresponding section in chap. 3 for calculation details).

Conservators' Note: *Laser digitization is used to gain information about surface topography; it probably would not be used on two-dimensional art, such as prints or textiles. Three-dimensional museum objects with low surface concentrations of colorants sensitive to red light probably should not be subjected to this technique, however.*

Miscellaneous Objects and Combinations of Materials

Asian lacquerware

The damaging effect of light on Asian lacquerware (urushi) is frequently mentioned (e.g., Umney 1987; several contributions to Bromelle and Smith 1988), but there is little detailed information on this in the Western literature. The need for repeated restoration of outdoor lacquered wooden structures has been cited (Hasegawa 1988). Japanese researchers (Kenzo 1986; Toyoshima 1991) have documented the deterioration of urushi when it is exposed to ultraviolet light from fluorescent black lights. Umney also has noted that lacquerware extensively exposed to light will become hydrophilic, that is, that some urushi resin will become soluble in polar solvents normally used for cleaning lacquer. Kumanotani (1988) has recommended that light of wavelength less than 365 nm be completely removed from exhibition lighting for lacquerware. These observations and recommendations are consistent with the fact that oligomeric urushiol, a component of Japanese lacquer, absorbs light below 360 nm. Kenjo (1988) has commented on the fact that extracts from Japanese cedar trees can be used in combination with urushi sap to conserve lacquered wood that has been exposed to light. The author notes that the color of these cedar extracts increases on light exposure. This possibility should be kept in mind when light exposure of repaired lacquerware is planned.

Webb (1997) undertook a study to delineate the effects of ultraviolet and visible light on lacquer objects. She found that one week of ultraviolet light of wavelengths below 360 nm was sufficient to cause solubility of some of the resins in Asian lacquerware in almost all solvents used for cleaning, although appearance changes were not noticeable at that stage. Four weeks of exposure were required to cause complete surface degradation. In another experiment, lacquer samples were continuously exposed to 6×10^3 lux of visible light until appearance changes were noticeable. After 100 days, before any changes were visible, solubility changes began to occur. The dose of light used in this experiment was very high, however, compared to the output from xenon flashlamps (6×10^3 lux for 100 days vs. 700 lux for 0.01 second from a point-and-shoot camera flash). On the basis of extensive study, Webb (2000) has assigned Asian lacquer a light sensitivity equivalent to Blue Wool no. 4, provided that ultraviolet light is rigorously excluded.

> **Conservators' Note:** *It can be concluded that lacquerware may be safely photographed with xenon flashlamps if ultraviolet wavelengths below 365 nm are rigorously excluded. Ultraviolet photography of lacquerware that has already received extensive light exposure is likely to pose considerable risk.*

Ethnographic objects

These objects can be made from a wide variety of natural materials or combinations of materials. The sections on fibers, papers, and natural dyes in this chapter are all relevant. Ethnographic objects could also include, for example, leather or hides, ivory and bone, feathers, fur, and colorful insect parts such as beetle or butterfly wings. Bradley and Daniels (1990) have summarized recommended display lighting levels for these objects, assuming they consist of known, rather fugitive materials or materials of unknown lightfastness. They reiterate the general rule of thumb presented years ago (e.g., Giles and McKay 1963) that materials absorbing in the blue are likely to be more sensitive to light than those absorbing longer wavelengths of light. The greatest concern is expressed for fugitive colorants (see the section on organic colorants).

Very little published experimental work on the light sensitivity of ethnographic objects was located. Horie (1990) investigated the fading of budgerigar feathers—in gray and bright shades of blue and green—from domestic caged birds. The data presented suggest that these feathers could withstand a total of up to several times 10^5 flashes from a point-and-shoot camera with ultraviolet blocking before a significant portion of their display lifetime is consumed (see corresponding section in chap. 3 for more details). Horie also states that an initial rapid change was observed, which increased if the light source contained near-ultraviolet light.

> **Conservators' Note:** *Horie's observations suggest that flash exposure of an object with blue or green feathers should be kept lower than 10^5 flashes for its entire*

display lifetime, particularly if near-ultraviolet wavelengths may be present.

Anecdotal evidence for the light sensitivity of the colored materials in insect collections exists, but no published data on this topic was found during the literature search.

Painted and pigmented clays

Montmorillonite and kaolin clays are recognized as being good surface catalysts (Thomas 1988). Since these and other clays are being used in increasing amounts in the manufacture of papers for various printing processes and in boards, the possible role of these substances in photoinduced degradation of art and archival materials may become important. Oven-dried clays are good adsorbers of aromatic molecules. For example, anthracenes are easily adsorbed onto and subsequently photodegraded on certain clays. The catalytic properties of the clays could have ramifications for the light exposure of certain colorants on papers containing clay fillers or coatings, as well as for that of painted clay objects.

Antimicrobial agents

Daniels and Boyd (1986) reported that crystals of thymol yellowed within a few days of exposure to light in a Microscal lightfastness tester when the light source was operated at 2×10^4 lux. Furthermore, Whatman no. 1 paper enclosed in a glass tube with the thymol crystals became visibly discolored after one week.

Conservators' Note: *Because this light exposure is many orders of magnitude greater than that from flashlamps or photocopier sources, the danger of discoloration of thymol-treated, paper-based objects due to flash photography is probably negligible. These experimental results, however, do indicate the possibility that deleterious changes may occur in thymol-treated objects that are kept in enclosed spaces and subjected to high light exposures.*

Johnson and Reagan (1990) tested the effect of six different commercial antimicrobial preparations on the fading of nylon fibers dyed with azo and anthraquinone colorants. These compounds are used to prevent mold growth in carpeting. The researchers found that four of the formulations did not have a significant effect on dye fading induced by exposure in an Atlas Fade-Ometer. However, two preparations—a mixture of phenolics and a mixture of a quarternary ammonium salt and an organotin compound—increased the rate and extent of dye fading. Interestingly, organotin alone did not have a deleterious effect. The authors speculated that interactions of the phenyl groups in the additives played a role in increasing light sensitivity. The data provided are for long exposures to light including near-ultraviolet wavelengths and cannot be used for calculating flash exposures; however, there is a possibility that this type of additive can increase the light sensitivity of dyed nylon.

Notes

1. Mathematical standards for expressing color units, issued by the Commission Internationale de l'Eclairage (International Commission on Illumination).
2. A process developed in 1891 by French physicist Gabriel Lippmann (1845–1921).

Chapter 3

Technical Details from the Literature Search

Colorants

Inorganic pigments

In Feller's 1968 study of vermilion, less than 100 hours of exposure to radiation from a carbon arc caused a significant change in the sensitive pigment. It is assumed for the purposes of this evaluation that this change is unacceptable and therefore that 100 hours of carbon-arc exposure represents complete use of the total display lifetime of the pigment. If this is the case, it is further assumed that only 0.01 of the equivalent exposure due to flash photography would be acceptable. An Atlas Fade-Ometer carbon arc emits about 0.02 W/cm^2 of energy in the near-ultraviolet region (McKellar and Allen 1979, 267); 0.016 W/cm^2 of this is in a very strong band centered at 395 nm (see fig. 1.8). The band at 450 nm is roughly 0.004 W/cm^2. These wavelengths are likely to be absorbed by vermilion and thus be responsible for the photochemical changes observed. In one hour (0.01 of the exposure that caused the unacceptable change) approximately 1.1×10^{20} photons/cm^2 in the 395 nm band are emitted by the carbon arc. This is arrived at by the following two steps:

1. The total energy of the photons emitted per square centimeter at that wavelength during the one-hour time period is calculated from the power as follows: total energy/cm^2 = 0.016 W/cm^2 $\times$ time = 0.016 J/s/cm^2 $\times$ 3600 seconds = 57.6 J/cm^2.
2. The number of photons per square centimeter is then calculated from equation 1.3 (see chap. 1), as follows: photons/cm^2 = 57.6 J/cm^2 $\div$ $(2 \times 10^{-16} \div 395)$ J/photon = 1.1×10^{20} photons/cm^2.

The same procedure shows that the allowable number of photons of blue light at 450 nm per square centimeter of object would be 3.2×10^{19}. Calculations of this type are used throughout this chapter.

Before determining the number of flashes that would irradiate an object with these amounts of photons, the effect of distance from the carbon arc to the sample should be considered. The data quoted by McKellar and Allen specify total emitted energies; it is assumed that this is at the surface of the light source. In one model of the Atlas Fade-Ometer, the sample distance was approximately 25 cm (10 in.); the irradiance at the sample would be 625 times less than at the surface of the light source because the light falls off inversely with the square of the distance. So the numbers of photons per square centimeter for comparison become about 2×10^{17} and 5×10^{16} for the 395 and 450 nm bands, respectively. The corresponding numbers of photons per square centimeter for a studio xenon strobe are about 2×10^{13} and 3×10^{13}, provided that the ultraviolet blocking does not effectively extend to 400 nm. The numbers would be about an order of magnitude less for a typical point-and-shoot camera. Comparison of these numbers suggests that 10^3 to 10^4 flashes from a xenon strobe would not adversely affect poor-quality vermilion and that 10^4 to 10^5 flashes from a point-and-shoot camera might be tolerated. Because the effect is cumulative, however, some limitation of flash photography may be prudent if the object is known to include poor-quality vermilion.

Natural organic colorants

Many natural red dyes are derivatives of anthraquinones or anthraquinone glycosides or consist of a few fused five- and six-membered rings with an extensive conjugated double-bond system (see fig. 2.1 for core dye structures of some common classes of colorants). Natural yellow dyes are more likely to be derivatives of flavonoids, which also have conjugated double bonds. The red and yellow colorants in resins consist typically of three or four fused rings, one or more of which is a heterocycle (i.e., a carbon atom in the ring is replaced by another atom, frequently an oxygen atom in the case of resins). An exception to these classifications are the yellow and orange carotenoid pigments, which are long polyisoprenoid molecules that in many cases have ring structures at the ends. The larger the extent of the conjugated ring system, the longer the wavelength of the absorption band in these compounds.

The natural blue dyes on museum objects are frequently indigoid dyes. These compounds are diindole derivatives that have mirror symmetry about a central double bond that connects the two indole moieties. Like the red and yellow colorants, they have a conjugated double-bond system. Tyrian purple is a dibromoindigo derivative. Purple, pink, and green organic colorants are most commonly mixtures of blue with red or yellow natural dyes.

Some blue dyes found in Japanese woodblock prints are anthocyanins. These are glucosides of flavonoid derivatives, and they may contain other covalently bonded aromatic moieties as well. When these dyes are solubilized from flowers, their colors will depend on the solution pH. In nature, anthocyanins probably form very specific aggregates that also involve metal ions. The aggregation and metal complexing appear to determine their colors in the plants (Goto and Kondo 1991). These researchers and their colleagues have succeeded in reproducing the aggregates and blue colors in the laboratory. It is possible that the colors of anthocyanin pigments in art objects are also dependent on the state of aggregation of the colorants.

Most brown dyes are tannins, a class of compounds that includes a variety of substituted polyphenols; in some cases, the phenol groups are connected by ester bonds. Many tannins are acidic. They are also present in leathers and in iron gall inks as complexes with iron. Another brown dye, juglone, is 5-hydroxy-naphthoquinone. Some other natural dyes can be made to look brown—or almost black—on certain fabrics by the appropriate choice of mordant.

Natural Japanese colorants

Feller, Curran, and Bailie (1984) exposed *aigami* samples to a total of 5.3×10^5 lm/m^2 of what they describe as soft-white fluorescent light during the two-day exposure. If the spectral distribution of soft-white fluorescent light is similar to that of cool-white fluorescent lamps (see "Fluorescent lamps," pages 21–23), then the *aigami* samples received a total of approximately 2×10^{16} photons/cm^2 of ultraviolet light between 320 and 400 nm, and approximately 3×10^{17} photons/cm^2 of visible light between 500 and 650 nm (which coincides with the visible absorption band of this dye). One xenon strobe flash will radiate about 2.5×10^{13} and 3×10^{14} photons/cm^2 of ultraviolet and visible light, respectively. Thus, about 10^3 flashes from a studio strobe setup could cause a perceptible fade in this very sensitive colorant.

Natural red and yellow Western colorants

Saunders and Kirby (1994a) presented extensive details for the fading of yellow buckthorn lake, which absorbs near-ultraviolet and blue light most effectively. Nine hundred hours of exposure to light from daylight fluorescent lamps filtered through a 400 ± 25 nm band-pass filter resulted in noticeable fade (ΔE ca. 5). In that time, the sample would have received approximately 6×10^{20} photons/cm^2 of light between 400 and 450 nm, and about one-quarter as many near-ultraviolet photons (375 to 400 nm). Because this amount of light caused a change greater than that which determines the total display lifetime for this pale, light-sensitive colorant, flashlamp exposure should probably be limited to less than 0.001 of this amount, or less than 5×10^{17} blue and 1×10^{17} near-ultraviolet photons/cm^2. This is about four orders of magnitude greater than a studio xenon strobe light will emit, so up to about 10^4 flashes could be tolerated, compared to fewer than 10^3 flashes for litmus and related dyes. The less sensitive colorants studied in the Saunders and Kirby research were indigo, quercitin, gamboge, and alizarin, as well as lakes of weld, lac, and madder.

In a study by Johnston-Feller and coworkers (1984), the Atlas Fade-Ometer xenon arc was filtered to simulate sunlight through window glass. The irradiation of the samples was monitored by measurement of the energy delivered at 420 nm (rather than as time of exposure) to eliminate uncertainties due to variations in lamp output. It should be remembered, however, that the irradiation included near-ultraviolet wavelengths above about 320 nm. To relate the data obtained in this study to possible effects of photoflash exposure, the spectral power distribution curves for the Fade-Ometer arc and a photoflash lamp are assumed to be the same. Approximately 20% of the colorant was lost in roughly 50 hours, during which time the samples received 74 kJ of light energy per square meter (or

7.4 J/cm^2) at 420 nm. If it is assumed that a 5% loss would be perceptible and would represent the total display lifetime of the object, then one-quarter of the experimental exposure would be equivalent to the dose during the total display lifetime. If it is also assumed that photoflash should deliver no more than 0.01 of this amount, then the approximate total allowable energy exposure at 420 nm from flash is calculated as follows: 7.4×10^3 J/cm^2 $\div 4 \times 0.01 = 0.019$ J/cm^2 (rounded to two significant figures). Application of equation 1.3 shows that this is equivalent to about 4×10^{16} photons/cm^2 at 420 nm.

A studio strobe emits about 3×10^{13} photons at 420 nm. If it is further assumed that the relative amounts of photons at the photochemically active wavelengths are the same for the experimental source and the flashlamps, then about 10^3 studio strobe flashes could be tolerated by alizarin lakes of the composition studied by Johnston-Feller and coworkers. At least an order of magnitude more flashes from point-and-shoot cameras could be permitted. These numbers may be overly conservative, however, because in actual use the Atlas Fade-Ometer source emits more near-ultraviolet light than the flashlamps. Yet the suggestion of a significantly greater initial rate of change when the samples were first irradiated (the steep initial slope shown in Johnston-Feller 1984, fig. 4) suggests that *pristine* alizarin lakes may begin to show changes after exposure to less than 10^{16} photons/cm^2. From this conservative point of view, about 10^3 flashes may be the maximum allowable for an object that has never before been displayed, and some restriction of photography may be desirable.

Green dyes, porphrins, carotenoids

Two recent reports on studies of other natural pigments contain information that may be useful. Quantum yields for photodegradation by visible light of hematoporphyrin in solvents of various polarities were determined by Móger, Köhler, and Getoff (1996). Hematoporphyrin is the iron-containing porphyrin derivative that is the chromophore in cytochrome *c* compounds, which are important electron-transferring substances involved in oxidation–reduction reactions in plants and animals. Their chromophores are similar to those of porphyrin chromophores in blood and to those of chlorophyll-based green pigments. Móger and coworkers found that the quantum yield for photodegradation of the hematoporphyrin is strongly dependent on the polarity of the medium. For a wavelength of 560 ± 20 nm and a nonpolar environment, it is greater than 0.001. The second report, on a study of the kinetics of photobleaching beta-carotene in solution, contains observations of a long-lived excited triplet state that was induced by laser flash at 355 or 532 nm, both wavelengths that are strongly absorbed by this pigment (Mortensen and Skibsted 1996). The effects were observed when chloroform, but not benzene, was used as the solvent. The authors remark that they had to change the sample every thirty laser flashes because of degradation. From the data they supply for the laser and for the pigment concentration in solution, it is calculated that when the sample has been irradiated with about two hundred times as many photons as there are beta-carotene molecules in the light path, it has deteriorated too much for continued use. This suggests that the quantum yield for degradation of the carotenoid used in the study was about 0.005. The role of the chlorinated solvent in the degradation of the colorant was not addressed, but it may have been a factor.

Synthetic organic colorants

Anthraquinone dyes

In the flash photolysis study of substituted anthraquinone dyes by Allen and coworkers (1981), the light source was a high-powered xenon arc with an output of 450 J per flash. Because the authors do not specify the distance of the sample from the source, or the extent to which ultraviolet light was blocked by materials in the optical path, the number and energy of photons reaching the sample cannot be computed. It should probably be assumed that ultraviolet light of wavelengths greater than or equal to 320 nm was active. A rough approximation suggests that between 2×10^{18} and 2×10^{22} photons per flash could have reached the sample. Typically there were about 2×10^{17} dye molecules/cm^3 of solution, and one (or only a few) flashes was enough to cause a measurable fading of the dyes. Thus quantum yields could have ranged from 5×10^{-5} to about 0.1.

Oda and Kitao (1985) and Kuramoto and Kitao (1981) irradiated anthraquinone, diphenylmethine, and indigoid dye solutions with a mercury arc placed 4 cm from the samples. The blue and ultraviolet wavelengths were blocked by a potassium dichromate solution used as a filter. The least-sensitive dye—the diphenylmethine compound—faded 15% in 60 minutes. In the same amount of time, the anthraquinone dyes faded 70–95%, and the indigoid dye about 40%. A quantum yield for fading of about 0.1 is obtained from their data, if the output of their high-pressure mercury arc is assumed to be equivalent to that of similar General Electric lamps in the appropriate wavelength range.

Other synthetic dyes

The light source in the microfading apparatus of Whitmore, Pan, and Bailie (1999) is a xenon arc that irradiated their gouache samples with 6.4×10^6 lux (or lumens/m^2) of visible light. A 20-second exposure to this light caused an unacceptable amount of fading (ΔE of 5) in the rose tyrien sample, so 0.01 of this exposure, or 0.2 second, would be the total allowable due to photoflash. This is equivalent to a total light dose of $0.2 \times 6.4 \times 10^6 = 12.8 \times 10^5$ lux seconds. The handheld flash camera discussed in chapter 1 delivers 700 lux per flash of 0.01 second, or 7 lux seconds of visible light energy per flash. Thus $12.8 \times 10^5 \div 7$, or about 2×10^5, flashes would be the maximum allowable over the lifetime of an object containing rose tyrien gouache. Rose bengal and magenta gouaches are about two-thirds as sensitive, so they could be subjected to about 3×10^5 flashes from a handheld camera, or about 3×10^4 flashes from a studio strobe setup with ultraviolet wavelengths blocked, during their display lifetime. It should be noted that these results are in qualitative agreement with the original findings of Hanlan (1970), who included rose bengal in his flash experiment.

Natural Fibers

Cotton

The shortest light exposure for which Buschle-Diller and Zeronian (1993a) present data is 168 hours. This amount of irradiation caused slightly less than a 10% decrease in tensile strength in the oxidized yarns that were more susceptible to damage. During this exposure time, which was equivalent to about 6×10^5 seconds, each square centimeter of material would have received approximately 5×10^{21} photons of near-ultraviolet light and at least half as many photons of blue light. (These numbers were calculated from data in the manual for the Atlas Weather-Ometer, after adjusting for a xenon lamp power of 3500 W instead of 2500 W, and for borosilicate filters as specified by the authors.) In contrast, one flash from a handheld camera delivers on the order of 10^{13} near-ultraviolet and blue photons/cm^2 in about 0.01 second (see table 1.2). Thus is would take about $5 \times 10^{21} \div 10^{13} = 5 \times 10^8$ camera flashes to cause the smallest change reported in the cotton samples. Assuming that flashlamp exposure is to contribute no more than 0.01 of the total detectable light damage over the display lifetime of the fabric, more than 10^6 flashes could be allowed.

Silk

Becker and Tuross (1994) used a total dose of 100 kJ/m^2 of visible plus near-ultraviolet light to cause a loss of about 10% of the tyrosine in the silk habutae test fabric placed in the Atlas Weather-Ometer. However, the filter used on the xenon source allowed ultraviolet wavelengths as low as 310 nm to irradiate the fabric. The authors do not indicate the relative effectiveness of the wavelengths below 365 nm. The energy dose they used may correspond to approximately 2.5×10^{19} photons in the visible region and 8×10^{18} photons in ultraviolet light between 310 and 400 nm. An estimated total of about 2×10^{18} ultraviolet photons/cm^2 could be allowed from flash assuming that (1) the same spectral distribution is emitted by the xenon flashlamps (this might result in a significant overestimate of the risk due to photoflash), (2) flash should cause no more than 0.01 of the maximum tolerable change, and (3) this would correspond to a loss of about 5% of the tyrosine. Data in table 1.2 show that fewer than 10^4 strobe flash exposures, or more than 10^5 flashes of a handheld camera, could be allowed over the lifetime of this type of undyed silk.

Kelly, Mollah, and Wilkinson (1990) irradiated a 2 cm^2 area of raw silk fabric with a 20-nanosecond laser flash containing approximately 10^{17} photons of 354 nm near-ultraviolet light. This induced a transient event, from which photochemistry is presumed to have proceeded, that lasted approximately 3×10^{-6} second. In contrast, a flash from the xenon studio flashlamp described in chapter 1 would irradiate the fabric with 10^{-4} to 10^{-3} as many photons as each laser flash. The photons would reach the silk over a 0.01-second interval, which is much longer than the excited-state lifetime of the active silk component. Consequently, only a very small percentage of the molecules would be excited at any one time by a xenon flashlamp, compared to the number affected by laser flash.

Pulp, Paper, and Wood

Cellulose and rag papers

Lee, Bogaard, and Feller (1994) observed slight bleaching of artificially discolored, unsized alpha-cellulose paper after three days of exposure to visible wavelengths above 460 nm from a daylight fluorescent lamp with a cutoff filter. The irradiance from the light source in this part of the visible spectrum can be from one-half to two-thirds of that of a xenon arc; it is also somewhat similar in spectral distribution. Thus the relative effectiveness of the experimental light source can be compared to a photoflash by comparing the exposure times. Three days in the fading apparatus is equivalent to 2.6×10^5 seconds, whereas the flash lasts 0.01 second. The ratio is approximately 3×10^7, suggesting that 10^7 to 10^8 flashes would be required to cause the reported slight bleaching, which is a just-perceptible change. If flash exposures are to cause no more than 1% of this change, approximately 3×10^5 flashes could be allowed over the display lifetime of the object.

The details supplied by Whitmore and Bogaard (1994) permit calculation of the dose in photons per square centimeter from each light source that caused unacceptable changes in their samples. When the power outputs of the lamps are converted to approximate rates of photons per square centimeter, the irradiation by the ultraviolet-A lamp is found to be approximately 4.5×10^{13} photons/cm^2/0.01 second, which is similar to flashes from two studio xenon strobes without color correction. The ultraviolet component of the daylight fluorescent lamp is 0.05 of this amount, or approximately 2×10^{12} photons/cm^2/0.01 second. These quantities have been multiplied by the exposure times to obtain the abscissa values for fig. 2.3. The visible component of this light source, rated at 1.8×10^4 lux, appears to cause about the same amount of deterioration of the paper samples as does the ultraviolet component (see fig. 2.3, compared to fig. 2.2). The visible output of a single flash from a studio xenon flashlamp is about equivalent to this amount or even more, depending on filtering, although the spectral distribution is different (compare fig. 1.3a and fig. 1.6). The changes Whitmore and Bogaard observed after the test papers were exposed to 74 hours of light from the daylight fluorescent source would have been more than what would occur during the entire display lifetime of the paper. Because the output per 0.01 second (the flash duration) of this source is about the same order of magnitude as that of a xenon flashlamp, the conclusion can be drawn that roughly 10^7 to 2×10^7 flashes from the flashlamp would be required to cause similar, unacceptable change in rag papers.

Thus approximately 10^5 flashes from the studio strobe and more than 10^6 flashes from a handheld camera with ultraviolet blocking would be the acceptable maximum number of flash exposures.

Lignin-containing papers

In a study of the photochemistry of lignin model compounds, twenty laser pulses at 355 nm were sufficient to cause a significant change in alkoxyphenols dissolved in organic solvents (Shkrob, Depew, and Wan 1992). The quantum yield for photodegradation of methoxy-substituted aromatic ketones was increased by an order of magnitude—to as high as 0.5—if the solvent contained alcohol or water, for example, rather than being nonpolar and unable to participate in hydrogen bonding. An increase in yield was not observed in the presence of a free radical scavenger, suggesting that free radicals are involved in the photochemical reactions (Schmidt et al. 1993). In another study, methoxy-substituted quinones incorporated into hydroxypropyl cellulose films or deposited onto filter papers were reduced to a significant extent by 30 minutes of radiation above 300 nm in a xenon fading apparatus (Castellan et al. 1993). These studies used light sources with a major ultraviolet component. Although they help to elucidate the nature of the photochemistry involved in lignin degradation, they cannot be used numerically to determine flash exposures.

Laser flash photolysis research on thermomechanical pulps indicates that exposure to flashes of approximately 10^{18} photons of 354 nm light can result in significant changes in the reflectance and absorption spectra of the pulp (Schmidt et al. 1990; Schmidt et al. 1991; Wilkinson, Goodwin, and Worrall 1993). The intermediates that formed following excitation are relatively long-lived, and the quantum yield for formation of yellow chromophores is about 0.004. For a paper made of this pulp and containing about 10^{19} monomer units of cellulose per square centimeter (see chap. 1), and a xenon photocopier light source such as that described in table 1.2, the fraction of monomer units altered would be equal to the quantum yield times the number of photons/cm^2 divided by the total number of monomer units/cm^2. This equals 0.004×10^{15} photons/cm^2 $\div$ 10^{19} monomer units/cm^2 $= 4 \times 10^{-7}$. That is, one monomer unit in 2.5×10^6 is photochemically altered. If a change in 2% of the monomer units is detectable, and if copying should cause no more than 0.01 of this change, then a maximum of 500 copies could be allowed. However, a moderate number of xenon camera flashes, which have a much lower ultraviolet component, would not be expected to pose a significant threat.

In experiments by Tylli, Forsskåhl, and Olkkonen (1995) on unbleached and bleached pulps, the ozone-bleached pulp began to show measurable changes within 15 minutes of irradiation. The instrument probably had a xenon-arc light source.[1] The wavelength used for the research is lower than those emitted by filtered photographic light sources, so moderate amounts of flash photography would probably not be deleterious. However, photocopying is probably not benign. Fifteen minutes of irradiation from a xenon source would have subjected the sample to approximately 10^{19} photons/cm^2 of 350 nm light. For comparison, the output of a xenon photocopier light source at 350 nm is roughly 2×10^{14} photons/cm^2/copy. If the noticeable changes that were observed experimentally are assumed to be a hundred times the change permitted due to reprography, then approximately 10^3 copies would be the maximum allowable over the lifetime of a document on ozone-bleached, groundwood pulp paper.

Ek and coworkers (1993) monitored the brightness of both bleached and unbleached groundwood pulp at 457 nm to determine how this material responded to radiation containing near-ultraviolet and visible light. The exposures were followed either by dark storage or by exposure to the equivalent of indoor daylight lighting. The investigators found that yellowing was partially reversed during storage, but that only 5 more minutes of exposure to their experimental light source after storage would cause reyellowing. The cycle could be repeated several times, with the pulp becoming more yellowed after each experimental exposure, as if it had been constantly exposed to the ultraviolet-plus-visible light source. Reduction of the pulp with sodium borohydride did not alter the response, but ascorbate did reduce the magnitude of the effect. This suggests that free radicals, but not carbonyl or aldehyde groups, are involved in the chemical reactions that result in yellowing.

Robert and Daneault (1995) investigated the yellowing of thermomechanical pulp handsheets by near-ultraviolet light from a broadband source with a maximum at 350 nm. This source emits light up to approximately 410 nm. They detected small changes in papers receiving a total irradiance of 3×10^{19} photons/cm^2, a dose that may be equivalent to about 10^6 flashes from a studio strobe without near-ultraviolet blocking or color correction. The changes detected appear to have been less than what would be defined as the total display lifetime. The data suggest, roughly, that on the order of 10^5 flashes from a studio strobe setup could be allowed; that is, moderate flash photography of papers made from this pulp might be permitted.

Forsskåhl and Tylli (1993) have shown that a significant increase in the 450 nm reflectance of handsheets made from chemimechanical pulp, accompanied by a decrease in the near-ultraviolet reflectance, can be caused by only 2 hours of radiation at 390 or 410 nm at an irradiance of approximately 1.3 mW/cm^2. This is equivalent to exposing the paper to roughly 10^{19} photons/cm^2 at these wavelengths. For photoflash to cause no more than 0.01 of the reported change, exposures should be kept under 10^{17} photons/cm^2. This dose would be reached with fewer than 10^4 exposures from a xenon studio strobe setup; with approximately 10^5 flashes from a handheld camera that has typical ultraviolet blocking but is not blue-light color corrected; or by fewer than a hundred photocopies.

The photocycling experiments of Forsskåhl and Maunier (1993) were performed on handsheets made from bleached and unbleached thermomechanical pulps. The researchers demonstrated that 2 hours of monochromatic radiation at various near-ultraviolet and visible wavelengths caused noticeable changes in the reflectance spectrum of the sheets. Pertinent here is the observation that light of wavelengths 370, 470, and 650 nm, as well as other ultraviolet wavelengths, caused significant changes in the unbleached-pulp paper; and that light of wavelengths 390, 450, and 500 nm caused significant changes in the bleached-pulp paper. The light source was a short-arc xenon lamp; the relative intensity of the output at wavelengths above approximately 500 nm is one-third to one-fourth of the intensity in the 370–440 nm range for this source. According to the authors, the spectral irradiance of the light delivered by the monochromator at wavelengths from 370 to 440 nm was approximately 1.3 mW/cm^2. These numbers suggest that roughly 7×10^{18} photons of blue and ultraviolet light, or approximately 10^{20} photons of red light, caused easily noticeable changes in the appearance of these papers. After initially exposing the papers to ultraviolet light of 290 nm, the cycling experiment was begun by exposing the samples to, for example, 45 minutes of blue light (above 430 nm) and then 45 minutes of near-ultraviolet light (370 nm). During this treatment, the papers showed gradual yellowing. The cycle could be repeated approximately a dozen times before the longer-wavelength light was unable to decrease the yellowness. The data from this research suggest flash and photocopy exposure numbers similar to those calculated for the experiments described in the previous paragraph.

In their earlier experiments, Lee, Bogaard, and Feller (1989; see also discussion of their work under "Cellulose and rag papers," pages 132–33) exposed a variety of handsheets and aged filter paper to approximately 5.5×10^{20} near-ultraviolet photons/cm^2. In the 1994 study, these investigators exposed Kraft paper to 8 hours of light from which the ultraviolet wavelengths had been removed. This exposure was about 0.1 as great as that used in the earlier experiment, in which slight changes occurred in the filter paper. This suggests that approximately 2×10^6 flashes might cause noticeable changes in Kraft papers. Total flash exposure over the display lifetime of objects on Kraft paper supports might best be restricted to approximately 2×10^4 flashes.

Andrady and coworkers (1991) used a large spectrograph to irradiate paper samples with monochromatic light at specific wavelengths between 280 and 600 nm. In one experiment, fresh newsprint irradiated with 5.5×10^{16} photons/cm^2 of visible light of 400 or 600 nm showed small changes. A studio setup of four xenon strobe lamps might emit 10^{13} to 10^{14} photons at these two wavelengths. Thus, about 10^3 flashes would deliver a dose equivalent to that which caused a slight change in the newsprint in this experiment.

Andrady and Searle (1995) irradiated newsprint made from softwood thermomechanical pulp with different wavelength ranges of ultraviolet or visible light, using cutoff and band-pass filters. They observed maximum yellowing by ultraviolet light between 335 and 385 nm. Subsequent irradiation with blue light between 395 and 420 nm, or visible light at 500 nm or 600 nm, caused photobleaching of the yellowed samples, as detected by a decrease in the yellowness index. The pH of the papers was lowered significantly by these treatments. If, instead, samples were first irradiated with "white light" (the authors do not specify whether this light included near-ultraviolet as well as all visible wavelengths), the subsequent responses to additional exposure to visible light of particular wavelength ranges were not reproducible. Calculations by the authors showed that the degradation of the newsprint was a linear function of the intensity of the light. Insufficient numerical data are provided to calculate recommended photoflash exposures, however.

The same authors also observed that when paper made from thermomechanical pulp was exposed to near-infrared light at elevated temperatures, yellowing would occur that was noticeably increased if green or red light was included with the near-infrared light. The extension of these results to photoflash phenomena has not been undertaken because it is assumed that flash exposures of objects in collections are conducted under conditions that avoid heating of the objects.

Wood

Dirckx and coworkers (1992) and Mazet and coworkers (1993) exposed wood samples to 365 nm ultraviolet light or to visible light from a Hanau mercury-vapor lamp. The energy of the visible output was greater than 60% of the total energy emitted by the source, but it was much less effective than the ultraviolet light. Significant changes in the wood were observed after 32 hours of irradiation, during which time the samples would have received approximately 7×10^{20} photons of near-ultraviolet light/cm^2 and well over 10^{21} photons/cm^2 of visible light. To keep the response to photoflash below 0.01 of the significant changes observed in this study, exposure to near-ultraviolet photons in the 360–370 nm range should be kept below approximately 10^{19} photons/cm^2. About 2×10^5 flashes from studio strobes would be the maximum allowable in this case, suggesting that moderate flash exposure would be permissible.

Gums and Natural Resins

De la Rie (1988a, 1988b, 1993) observed that 378 hours of irradiation in a xenon-arc fading device yellowed films of dammar resin significantly and that use of an ultraviolet-blocking filter reduced the yellowing by a factor greater than 5. If the changes observed in the experiments with the ultraviolet-blocking filter in place are considered to be similar to the changes that define the display lifetime of the resins, flash exposure should be kept to 0.01 of this, or to the equivalent of approximately 3.8 hours of fading device output in the visible. This amounts to approximately 10^{21} photons/cm^2, or to about 10^6 times the visible irradiation from a studio strobe flash with ultraviolet filtering.

Synthetic Polymers

Polyolefins

Bigger and Delatycki (1989) obtained a quantum yield of approximately 0.08 for oxygen uptake by unstabilized high-density polyethylene when the sample was irradiated with a xenon arc that delivered approximately 5×10^{13} photons/s/cm^2 at 340 nm. This means that approximately 4×10^{12} oxygen atoms were taken up per second by each square centimeter of polymer. The researchers did not indicate to what extent other wavelengths were filtered from the light source. In a later paper, they showed that the rate of polymer oxidation was dependent on the intensity of the light source and was also influenced by the presence and type of ultraviolet stabilizers (Bigger, Scheirs, and Delatycki 1992). Exposures of 100–1000 hours were required to induce changes in the bulk physical properties of the polymers. Xenon photoflash lamps probably radiate fewer than 0.001 as many photons at 340 nm as the xenon arc used in this experiment, and the absorbance of the polymer is significantly lower at longer wavelengths. Thus, it can be concluded that photoflash lamps will not induce sufficient oxidation of polyolefins to cause changes in their bulk physical properties. The presence of ultraviolet stabilizers would make the possibility of damaging levels of oxidation even more unlikely. This can be seen from the following analysis: The density of the experimental polyethylene samples was 0.954 g/cm^3, which is equivalent to approximately 6×10^{22} monomer units/cm^3. Consequently, about one monomer in 10^{10} was oxidized in the experiments. In the case of photoflash exposure, not even one monomer in 10^{12} would be altered per flash.

Similar conclusions can be drawn from the results of studies on other polyolefins, such as low-density polyethylene and polypropylene (Rånby and Rabek 1992; Rånby 1993). The presence of transition metal ions or titanium dioxide can increase the rate of oxidation and embrittlement of polypropylene by up to a factor of 6, depending on the manufacturing process and the concentration of these components (Allen and Edge 1992d). This increase would not be sufficient to alter the conclusions regarding exposure to photoflash sources, however. Allen's group has also shown that the more highly fluorescent a polyolefin is in ultraviolet light, the more rapidly it will be oxidized. Again, the wavelengths used are lower than those expected to be present in significant amounts in a xenon photoflash light source.

Antioxidants such as di-*tert*-butylhydroxytoluene (BHT) have been incorporated into polyolefins for years. As they scavenge oxidizing radicals, these protective compounds are transformed into degradation products, some of which may be colored. Kovářová and coworkers (1995) investigated the effects of model degradation products for antioxidants on the photooxidation of low-density polyethylene. They used an Osram fluorescent tube emitting wavelengths from 315 to 405 nm, but they do not specify the intensity of the irradiation. None of the antioxidants or degradation products they used acted as either photoinitiators or photostabilizers under the experimental conditions. However, a colored stilbene quinone degradation product was converted to a colorless compound during irradiation of the samples. Not enough experimental details are provided to permit calculations relevant to photoflash exposures, but the overall results suggest that the presence of antioxidants or their breakdown products in polyolefin objects probably would not significantly alter the sensitivity of the objects to photoflash sources.

Polyesters

Exposure of poly(ethylene terephthalate) to carbon-arc radiation for a few hundred hours caused as much or more change on the rear surface of a sample as did exposure to a xenon arc, although the latter caused greater changes in the front surface of the sample (McKellar and Allen 1979, 149). This observation could be due in part to formation of a degradation product on the front surface that absorbed additional wavelengths, preventing them from traveling deeper into the material. Details of the exposure were not provided, but the total exposure of 200 hours to the carbon arc would have exceeded 10^{22} photons/cm^2, or about 10^8 times the near-ultraviolet and blue photons emitted by a single xenon studio strobe flash.

In a study of poly(ethylene 2,6-naphthalate), Allen and McKellar (1978) observed that the spectrum of the polymer was markedly changed after only twenty flashes of a source that delivered approximately 9×10^{16} photons per flash at 347 nm. One and a half hours of exposure in a Microscal apparatus with a 500 W mercury–tungsten source had a similar effect. The latter would have delivered approximately 10^{19} photons/cm^2 in the near-ultraviolet and blue ranges during that time. If the change observed is considered unacceptable, flash exposure should contribute only 0.01 of this amount, or 10^{17} photons/cm^2. About 10^3 flashes from a studio strobe setup would do this.

Polystyrenes

Months of exposure of polystyrene foam to sunlight caused yellowing and embrittlement of the surface, with the thickness of the altered layer depending on the length of exposure (Pegram and Andrady 1991). Ninety hours of exposure to light of wavelengths greater than or equal to 300 nm caused oxidation of polystyrene films (Lemaire et al. 1995). Andrady (1997) determined the activation spectrum in the near ultraviolet for yellowing of extruded polystyrene foam sheets. Wavelengths above 340 nm appeared to be less than 3% as effective as shorter wavelengths in causing the yellowing. These active wavelengths are not a significant component of flashlamp output.

In their study of copolymers of styrene and acrylonitrile, Lemaire and coworkers (1995) found that the copolymer was more stable to photooxidation than the polystyrene alone. Jouan and Gardette (1991) observed the effects of exposure to light of wavelengths greater than 300 nm on blends of this copolymer and polybutadiene. (This mixture, called high-impact polystyrene, or HIPS, gains its resiliency from the elasticity of the butadiene polymer and its strength from the polystyrene.) They observed a significant increase in the carbonyl band in infrared spectra of samples irradiated for 17 hours by the visible output from a mercury-arc lamp. The total visible irradiance at the sample would have been about 90 W/m^2. If the difference in spectral distributions between the mercury-arc and xenon-arc lamps in the visible region is assumed to be small, it can be calculated that the 17-hour exposure is equivalent to more than 10^{10} flashes from xenon photoflash lamps. Flash exposure would not be expected to cause a problem.

Hoyle and coworkers (1994) investigated the light stability of copolymers of polystyrene and poly(phosphazenes), which are used outdoors as fire-retardant films and coatings. Irradiation at wavelengths below 300 nm failed to alter the absorption spectra of the material in the visible range, and ultraviolet light of wavelengths below 340 nm was required to cause oxidation and chain breaking. These results imply that the copolymers are unlikely to be affected by light sources used in photography and photocopying.

Pickett (1990) reviewed the photostability of blends of polystyrene and poly(phenylene oxide) resins. These blends are used to make molded cases for appliances such as computers. It is known that extensive exposure to indoor fluorescent lighting, as well as to sunlight, causes color changes in these materials. Using ultraviolet light of various wavelengths, Pickett determined that wavelengths below 320 nm discolor the copolymers and that blue light tends to photobleach the discolored resin. The 365 nm mercury line appears to be without effect. The photochemistry is complicated; two or more pathways are involved, and these are influenced to different extents by temperature. The kinetic data suggest, however, that approximately 10^{20} photons of blue light would be required to cause a significant change in the appearance of the resin blend. This is roughly 10^{6} times the output from a xenon strobe setup. Commercial copolymers of this type are now made with light stabilizers such as Tinuvin 770. The additives retard the photooxidation, so photoflash exposure can be presumed to be without significant risk.

Polyamides

Katsuda and coworkers (1996) used a large grating monochromator to spatially isolate light of specific wavelengths from a high-powered xenon arc in a manner similar to the separation of colors by a prism. They placed their samples at the different sites in the apparatus corresponding to the positions of particular wavelength bands. On the basis of the total irradiation energy stated in the report, and the spectral distribution published by Thomson (1986) for continuously operating xenon lamps, it can be approximated that more than 10^{22} photons in particular ultraviolet wavelength bands caused a ΔE of approximately 10 in the nylon samples. If a ΔE of about 1 could be detected in these undyed samples, then the number of photons per square centimeter that could be contributed by flash is 0.01 of 0.1 of the experimental dose, or 10^{19} photons/cm^2. This is equivalent to about 10^{5} xenon strobe flashes over the display lifetime of undyed nylon.

Anton (1965) showed that 20 hours of exposure to light between 340 and 415 nm caused a detectable decrease in tensile strength of undyed nylon-6,6 samples. Based on data for the Atlas Weather-Ometer xenon source, this exposure amounted to approximately 3×10^{21} near-ultraviolet plus blue photons/cm^2. This exposure is equivalent to more than 10^{7} xenon strobe flashes. Thus, over the display lifetime of a nylon-6,6 object, flash exposure should be limited to 0.01 of this amount, or to about 10^{5} studio strobe flashes. Similar numbers can be calculated from information published by Fornes's research group (Fornes et al. 1973;

Stowe et al. 1973), which exposed nylon-6,6 to light of wavelengths above 350 nm in a different apparatus.

Synthetic polyamides that contain phenyl groups (aromatic polyamides, also called polyaramids) can be somewhat more sensitive to near-ultraviolet radiation. The data obtained by Carlsson, Gan, and Wiles (1978) suggest that detectable changes in poly(phenylene isophthalamide) films will occur after absorption of 1.7×10^{20} photons/cm^2 of light at wavelengths above 365 nm. A studio xenon strobe flash emits approximately 6×10^{14} photons/cm^2 in that wavelength range per flash, so roughly 3×10^5 flashes would be unacceptably damaging. Although more sensitive than aliphatic polyamides, these polymers would not be endangered by modest exposure to flashlamps.

Johnson, Tincher, and Bach (1969) exposed films of various polyamides to light from a 900 W xenon arc dispersed by a large spectrograph. This source, before dispersion, can be assumed to have been similar to the strobe light used for studio photography. The wavelengths of maximum effect were between 360 and 370 nm. Four hours of exposure resulted in major changes in the poly(isophthalamide) of 4,4′-bis(*p*-aminophenyl)-2,2′-bithiazole (BTA). This is approximately 1.5×10^6 times as long as a single flash. Because this exposure caused major damage, it might be advisable to limit flash exposures to 0.001 of this—only about 10^3 flashes of a studio strobe lamp—to avoid detectable change. Other polyamides that do not contain thiazole components were also tested, and they were found to be hundreds of times less sensitive.

Polyacrylates and polyacrylics

Feller, Curran, and Bailie (1981), using a carbon-arc Atlas Fade-Ometer, showed that the rate of formation of insoluble material in poly(methyl methacrylate), when it is irradiated with visible light of wavelengths greater than or equal to 400 nm, is only 5% of that when ultraviolet light between 320 and 400 nm is included. In a comparison of poly(alkyl methacrylates) with different alkyl substituents, the 3-methylbutyl material was found to be over 10 times as sensitive as the 2-methylbutyl isomer. The polymers having alkyl substituents other than methyl were slower to cross-link than the 2-methylbutyl compound, and the commercial mixture of Acryloid B-72 was the most stable. Poly(3-methylbutyl methacrylate) became 50% insoluble after 4 hours in the Atlas Fade-Ometer when exposed to ultraviolet and visible light from the carbon arc. The authors report that they subsequently obtained similar results with a xenon-arc Atlas Weather-Ometer. If ultraviolet light is excluded, 80 hours would be required to cause the same extent of cross-linking. During this time, an object coated with this most sensitive varnish would have been irradiated with more than 1.4×10^{23} visible-light photons/cm^2. This is arrived at as follows: 80 hours $\times$ 3600 seconds/ hour $\times 5 \times 10^{17}$ photons/cm^2 $= 1.4 \times 10^{23}$ photons/cm^2. The extent of degradation observed could be as much as ten times the total display lifetime of the sensitive resin, so flash exposure should be kept to 0.1×0.01 of this amount, or 1.4×10^{20} photons/cm^2. Even for this most sensitive resin, the allowable exposure is about 10^5 flashes from a studio strobe setup.

Torikai, Hattori, and Eguchi (1995) reported wavelength-dependent studies of the photochemically induced breakdown of poly(methyl methacrylate). The incidence of chain scission was too small to be detected when visible light of 400 or 500 nm was the actinic source. The dose was approximately 4×10^{19} photons/cm^2 in these experiments. Deterioration was observed only if ultraviolet wavelengths of 340 nm or below were used. Andrady (1997) summarized work by Fukuda and Osawa, who found no measurable effect of irradiation at wavelengths between 265 and 500 nm from a mercury arc on poly(methyl methacrylate).

Poly(vinyl —) polymers

Martin and Tilley (1971) exposed samples of unstabilized poly(vinyl chloride) to a xenon arc in a weathering device, with the addition of a series of glass cutoff filters. They found, for example, that when wavelengths below 385 nm were blocked by a filter, the rate of oxidation (as measured by the formation of carbonyl groups) was only 6% of that observed when the entire xenon-arc spectrum was used; if the cutoff wavelength was 415 nm, less than 2% of the total amount of oxidation was detected. The researchers were also able to correlate the extent of yellowing of the polymer with the infrared absorption due to carbonyl groups. About 300 hours of radiation at wavelengths above 385 nm were required to induce formation of detectable amounts of carbonyls; the samples received more than 700 hours of visible radiation before measurable oxidation was detected. These results suggest that exposure of unstabilized PVC to xenon photoflash lamps with ultraviolet blocking, or to other light sources that radiate negligible energy at wavelengths below 385 nm, would probably not be deleterious.

Investigations by Andrady and coworkers of the photodegradation of rigid PVC (Andrady, Torikai, and Fueki 1989; Andrady, Fueki, and Torikai 1990) suggest that ultraviolet radiation becomes decreasingly effective in yellowing the polymer as the wavelength increases from 280 to 340 nm. Their data also indicate that the composition of the plastic product will influence its response to the ultraviolet and that blue light at 400–500 nm may cause very slight bleaching. Application of the equation published by Andrady (1997) for change in the Yellowness Index of PVC without any titanium dioxide shows that at wavelengths of 360 nm and 410 nm, 10^{21} and 10^{22} photons/cm^2, respectively, would be required to cause a perceptible change. This corresponds roughly to 10^7 flashes from a studio strobe setup. One percent of this would be approximately 10^5 flashes, which could be safely allowed.

Polymeric Materials Containing Colorants

Textiles containing natural dyes

Comparative studies of Western dyes on wool and cotton

Padfield and Landi (1966) exposed their samples to fluorescent lights with and without cutoff filters that block ultraviolet light below 400 nm or all near-ultraviolet and visible light below 560 nm. The results they presented for light without ultraviolet were all for a total of 3.6×10^7 lux hours of radiation, which is very approximately equivalent to 2×10^{21} visible-light photons/cm^2. This large dose caused major fading of turmeric and brazilwood mordanted with alum on wool and of brazilwood on cotton. Other samples that were affected to a lesser extent were cochineal, madder, and logwood mordanted with tin on wool, cutch on cotton, and quercitron, weld, and Persian berries mordanted with aluminum on wool. Brazilwood was the only dye tested that faded significantly when blue and green wavelengths, as well as ultraviolet, were removed from the light source. The shortest light exposure that Padfield and Landi used, when near-ultraviolet light was included, was about 6×10^6 lux hours. This exposure caused about as much damage as a sixfold larger visible-light dose to the turmeric sample, and about half as much fading of the brazilwood dyes as the longer exposure without ultraviolet did. The long irradiation times and the extent of fading do not allow specific conclusions to be drawn about the efficacy of permitting exposure of these particular colorants to photoflash lamps.

Gallotti (1994) included a brief list of results of his own fading studies on one natural dye for each color in the visible spectrum and with each lightfastness level on the Blue Wool scale from no. 1 to nos. 5 and 6. In order of increasing lightfastness, the dyes he listed were turmeric mordanted with aluminum; sandalwood with aluminum; safflower with aluminum; cochineal with aluminum; cochineal and logwood with aluminum and iron; madder with aluminum; cochineal with aluminum and tin; turmeric with aluminum, copper, and iron; and indigo.

Crews (1987) described light doses using AATCC fading units. Before her data can be evaluated, the AATCC unit needs to be stated in terms of the light energy radiated by a xenon arc. In this case, 20 AATCC units are equivalent to 3456 J/m^2. Thus, 1 AATCC fading unit equals 17.3 J/cm^2. In the experiments, exposure to 1 AATCC unit caused a ΔE of at least 10, which is a major fading. A ΔE between 1 and 2 should be perceptible for this yellow dye in a light shade on a textile, so 1 AATCC fading unit would consume more than the total display lifetime of the material. If photoflash is to be restricted to no more than 0.01 of the display lifetime, it should deliver no more than about 0.02 to 0.03 J/cm^2 of light energy. This is approximated as follows: $(1 \text{ to } 2) \div 10 \times 0.01 \times 17.3$ J/cm^2 = 0.0173 to 0.034 J/cm^2 (rounded to two significant figures). This much energy is roughly equivalent to 4×10^{16} near-ultraviolet and blue photons, or on the order of 100 flashes from a xenon studio strobe setup.

In the work of Crews (1988, 1989) on the effectiveness of several ultraviolet blocking filters, an exposure of 80 AATCC fading units was the only one for which fading data are given. This dose caused appearance changes with ΔE values of approximately 2.5–10 in the dyes other than turmeric when the amber filter, which blocked near-ultraviolet light plus a fraction of blue light between 400 and 550 nm, was used. This exposure is roughly equivalent to 2×10^{20} visible-light photons/cm^2. The change caused by this amount of light would be perceptible, so photoflash should provide no more than 0.01 of this amount over the lifetime of the object, or 2×10^{18} visible-light photons/cm^2. This dose could be delivered by fewer than 10^4 flashes of a xenon studio strobe setup *with filtering equivalent to the amber filter*. Filters in use with these sources probably transmit significantly more blue light than the amber filter used in this experiment, so exposure to xenon photoflash should probably be restricted to about 10^3 flashes total over the lifetime of the dyed textile.

In the experiments of Duff, Sinclair, and Stirling (1977), 80 hours of irradiation (the shortest time for which data are given) caused a change in b* of 10–25 units for wool dyed with sulfonated indigo, old fustic, and Persian berries. If an easily detectable change in b* for these light-hued yellow colorants is 2, and the fading obeys the reciprocity principle, then about 5 hour of irradiation from the Microscal lamp would represent the total allowable exposure. If flash is not to exceed 0.01 of this, the equivalent of 0.05 hour of Microscal radiation could be allowed. Based on the data provided by the researchers for the lamp output, this corresponds to about 4×10^{14} visible-light photons/cm^2, plus fewer than 4×10^{13} near-ultraviolet photons/cm^2. This limit might be reached in about 100 flashes from a filtered xenon flashlamp.

Red Japanese dyes on cotton

In experiments by Kenjo (1987) on red Japanese dyes on cotton, the total energy delivered at each of the wavelengths used to irradiate the samples was 8.5 J/cm^2. Use of these data with equation 1.3 shows that this energy corresponds to 1.4×10^{19} and 3×10^{19} photons/cm^2 for the shortest and longest wavelengths, respectively. An English-language summary of the resulting color changes in the samples suggests that approximately 2×10^{19} photons/cm^2 of blue light greatly exceeded the dose required to cause a perceptible change in the sappan-wood red and safflower dyes. If 0.01 of this number of photons is the approximate upper limit for flash exposure, then it can be concluded that about 10^3 flashes from a xenon photoflash lamp may represent the maximum allowable exposure for these red colorants on cotton.

Experimental studies on degraded natural dyes on textiles

Nasu, Nakazawa, and Kashiwagi (1985) irradiated several traditional yellow Japanese dyes—flavones, flavonols, an alkaloid, and a carotenoid—on silk with a xenon lamp in a fading apparatus for 40 hours, and observed ΔE values ranging from less than 5 to almost 20. If a total allowable ΔE is about 1 (a change that can be perceived by eye in these light yellow colorants), and flash exposure should cause no more than 0.01 of this, then 0.02 hour of xenon lamp radiation would be the maximum permitted. If the output from the lamp in the fading apparatus used by these investigators was roughly equivalent to that from a xenon-arc Atlas Fade-Ometer, then the allowable photoflash exposure would be a few hundred flashes of a xenon studio strobe without near-ultraviolet filtering. This suggests that several hundred such photoflashes might be the maximum allowable for materials containing these particular yellow colorants.

Textiles containing synthetic colorants

Anthraquinone and azo dyes

Allen, McKellar, and Protopapas (1977), studying a yellow anthraquinone derivative on nylon, observed almost 30% fading in air in 5 minutes in a Xenotest weathering instrument under conditions mimicking natural sunlight. If the output of the light source in this instrument was comparable to that in a modern apparatus, the samples received roughly 5×10^{17} near-ultraviolet photons between 320 and 400 nm per square centimeter, and approximately 10^{19} visible-light photons/cm^2. Assuming an acceptable amount of fading due to flash exposure would be 0.1% (equivalent to 0.01 of a perceptible 10% fade), then approximately 2×10^{15} near-ultraviolet and 3×10^{16} visible-light photons should be allowed. These levels of exposure would be reached in about 100 flashes from a studio strobe. However, it should be realized that this colorant may have been sensitive to the near-ultraviolet wavelengths below 360 nm emitted by the test source. These wavelengths are removed by filters from studio and camera flashlamp outputs. Thus the calculations may significantly overstate the possibility of fading.

In a study of the fading of C.I. Acid Red 1 and C.I. Acid Orange 8 on silk fibroin, Massafra and coworkers (1998) exposed the dyed fibers to a total output of 94 W of near-ultraviolet light plus visible light from a xenon arc in a Xenotest instrument. These dyes are more sensitive to blue than to near-ultraviolet radiation. The kinetic data presented suggest that the lifetime of the orange-dyed fibers under these conditions was between 20 and 40 minutes, during which time the samples would have been exposed to approximately 8×10^{16} photons of blue light/cm^2. If photoflash is to deliver no more than 0.01 of this, several hundred flashes from a studio xenon setup might be the maximum allowable over the lifetime of silk fabric dyed with C.I. Acid Orange 8. The red dye was more lightfast but appeared to cause tendering of the silk fibroin with similar kinetics. The lightfastness of these dyes is not markedly affected by pH in the range 3–9. The ongoing research by this group should continue to provide useful data in the future.

Mehta and Peters (1981, 1982) investigated the lightfastness of cellulose acetate and polyester fibers dyed with many different derivatives of substituted aminoazobenzene dyes. All of these compounds had absorption maxima in the blue in ethanol solution; that is, they are yellow dyes. The most light sensitive among them had stabilities on polyester fibers equivalent to Blue Wool no. 1. This result suggests by analogy that no more than 10^4 flashes might be the maximum allowable for polyester textiles dyed with the most sensitive of these colorants. However, not enough data are provided to perform a direct calculation.

The lightfastness of a variety of monoazo derivatives containing heterocyclic substituents was studied by Savarino and coworkers (1989). These colorants were used to dye nylon, cellulose acetate, and polyester; they were also compared to cationic derivatives of the dyes on acrylic textiles. The lightfastness was evaluated by comparison to Blue Wool standards in a Xenotest apparatus. Slight modifications in structure (e.g., moving a substituent to an adjacent carbon atom in a ring) could cause changes in lightfastness up to three grades on the Blue Wool standard scale; the fastness of the same dye on the different substrates frequently varied by two grades. Although most of the dye-fabric combinations had lightfastnesses ranging from Blue Wool no. 3 to no. 5 or higher, one compound on cellulose acetate was rated at Blue Wool no. 1. Also, two of the cationic dyes on acrylic—a thiazole and an oxazole monoazo derivative—were markedly more fugitive than the other dyes on this textile. Instead of providing numerical data, the authors describe the response of these colorants to light by stating that "under irradiation the color of the specimens is deeply impaired" (Savarino et al. 1989, 170). Lack of detailed information on the exposure conditions and the fact that 10–15% of the radiated energy was in the near-ultraviolet range make more specific analyses of these data impossible. However, the results do suggest that unlimited flash photography of the dye-fiber combinations studied would be unwise.

Carpignano and coworkers (1985) measured the lightfastness of forty-four azo dyes on polyester fabric and in polyester film. The values ranged from no. 1 to no. 7 on the Blue Wool scale. The fastnesses of individual compounds were similar in the two forms of the polymer. Mathematical correlations were obtained between some properties of the substituents on many of the dyes and the tendency to photodegrade. Apparently small structural differences among the colorant compounds led, in some cases, to very large differences in fastness ratings.

Baumgarte and Wegerle (1986a) reviewed several aspects of the lightfastness of disperse and reactive azo dyes. They comment that dihydroxymethyl urea-based, crease-resistant furnishes manufactured during the 1960s and 1970s lowered the lightfastness of azo dyes on fabrics containing those compounds. They also quote unpublished results by Krallman on reactive yellow azo dyes; those results indicated that the lightfastness of some dyes on damp cotton was orders of magnitude lower than it was on dry cotton. The lightfastness of other, similar yellow dyes was essentially unaffected by dampening the fabric; the responses of the dyed fibers to humidity changes were not similar with respect to light sensitivity. Unfortunately, the numerical data needed to calculate flash exposures are not provided in the review article.

Sirbiladze and coworkers (1990) used a Xenotest aging apparatus to compare the lightfastness of several related derivatives of reactive azo dyes on cotton. Minor changes in substituent location or structure resulted in as much as an order-of-magnitude change in lightfastness. The most sensitive compound studied had faded 10% in 7×10^3 seconds in the Xenotest, which suggests that it is approximately one-seventieth as sensitive as the anthraquinone derivative investigated by Allen and coworkers (1983) described earlier. If the same assumptions are used in this case—that is, that 0.01 of the discernable change would be the maximum allowable due to flash photography and that the Xenotest irradiation conditions were similar to those in an Atlas Weather-Ometer—then more than 10^4 flashes from a xenon strobe lamp without near-ultraviolet filtering could be allowed before the occurrence of unacceptable fading of these reactive azo dyes on cotton. More recently, some of these investigators presented data suggesting that the fading of similar azo dyes on cotton was accelerated approximately threefold in the presence of the finishing agent dihydroxymethyl ethylene urea (Vig et al. 1996). The published experimental details were not sufficient to permit calculation of allowable photoflash exposures for these systems.

Another group of heterocyclic monoazo dye derivatives with closely related structures were synthesized by Christie, Dryburgh, and Standring (1991), who investigated their fastness properties. Although color was improved by replacing carbocyclic moieties with the heterocyclic ones, lightfastness of the new pigments was markedly decreased. Possible flash exposures cannot be calculated from the data supplied in the report.

Other dyes on textiles

In experiments by Oger (1996), yellow, orange, red, wine, blue, and gray colorants were each used alone at 3% dyeings on linen; dyeings of mixed colorants were used to create plum (wine and gray dyes) and beige (orange, bronze, and chestnut dyes). The samples were aged in a Xenotest apparatus set up to simulate sunlight through glass, so near-ultraviolet wavelengths between 320 and 400 nm were included. The orange- and beige-colored fabrics that had been treated with a fixative (3% Tinofix) were the most sensitive, showing a ΔE of 3 after approximately 15 hours of exposure (the shortest time for which data are presented). Again invoking a maximum acceptable ΔE of 1 and allowing photoflash sources to contribute only 0.01 of this, on the order of 2×10^4 flashes from an unfiltered xenon strobe lamp might be the maximum allowed over the entire display lifetime of a linen fabric dyed with the orange dye Orange TGL.

Wilkinson and his colleagues (Wilkinson, Willsher, and Pritchard 1985; Wilkinson and Willsher 1987) used laser flashes at 532 nm or 354 nm to excite their textile samples. Data for 532 nm excitation of rose bengal on nylon permit calculation of an acceptable number of flash exposures. The laser light excites the dye molecules in the middle of an absorption band. The approximate quantum yield for molecules that do not recover from the flash and the approximate irradiance in this wavelength band from a xenon studio flashlamp were used to calculate that each strobe flash might photochemically alter between 10^{11} and 10^{12} molecules of this fluorescein derivative per square centimeter on nylon. For a moderate dyeing of 10^{16} colorant molecules/cm^2, approximately 100 to 10^3 flashes could lead to photodegradation of 1% of the dye; this might be a conservative maximum allowable change due to photoflash for the display lifetime of the object. Flash exposure of this particular dye-fiber combination should probably be restricted to necessary studio photography.

Phototendering of dyed cotton, polyamide fibers, and wool

In phototendering studies, Reinert (1988) found that, at 60% relative humidity, 600 hours of xenon light exposure with wavelengths less than or equal to 320 nm blocked, reduced the tear strength of dyed polyamide by approximately 55%. If a 1% reduction in strength is considered acceptable over the display lifetime of the material and the light was equivalent to an Atlas Weather-Ometer exposure, then approximately 10^{21} near-ultraviolet photons/cm^2 could be tolerated by the dyed polyamide, and flashlamp exposure should deliver no more than 0.01 of this amount. On this basis, polyamide fibers dyed with these types of phototendering dyes could be exposed to about 10^5 flashes from a xenon source that includes near-ultraviolet light and to at least an order of magnitude more flashes from handheld cameras.

Organic dyes and pigments in other polymeric materials

Dyes in acrylic, epoxy, or polyester resins

Tennent and Townsend (1984) irradiated acrylic, epoxy, or polyester resin samples containing solvent dyes or pigments. The samples were exposed in a Microscal apparatus with a lamp that emitted a significant portion of its radiation in the mercury 365 nm band. Their numerical results cannot be used in the current inquiry; however, several interesting observations were made. None of the dyes in the polymer samples were as fugitive as Blue Wool nos. 1 and 2. With the exception of C.I. Solvent Red 50, a xanthine dye, the colorants were much more stable in acrylic than in the other polymers. The authors suggest that the amine hardener in the epoxy contributed significantly to dye fading. C.I. Pigment Yellow 95, an azo colorant, protected the resins from yellowing, even at long exposure times, and was little affected itself. The anthraquinone dye C.I. Solvent Violet 13 continued to change color when stored in the dark after 11.5 hours of irradiation (the shortest time for which data are reported), as well as after longer exposure times. Of the dyes tested, only this one and the xanthine need be of concern to conservators because of possible sensitivities to the equivalent of moderate amounts of photoflash exposure.

The light source used by Allen and coworkers (1983) to investigate the photodegradation of three anthraquinone dyes in epoxy-resin films was the 500 W Microscal lamp. This source includes a small component of near-ultraviolet light below 360 nm, and approximately 10% of the radiated energy is in the mercury 365 nm band. However, the effects reported are unlikely to be due solely to the ultraviolet component of the radiation, as the following analysis shows. The epoxy film samples contained approximately 2.5×10^{19} dye molecules/cm^2, and the total irradiance during a 10-minute period was roughly 10^{19} visible-light photons/cm^2 plus approximately 10^{18} ultraviolet photons/cm^2. During this time, a few percent loss of visible absorption was observed for the most sensitive dye studied, which suggests that well over 10^{18} molecules reacted. Unless the quantum yield for the ultraviolet wavelengths was greater than unity (which is highly unlikely), the visible portion of the Microscal radiation must have contributed to the observed fading of the dye. The overall quantum yield, calculated from the approximate numbers given here, may have been as high as 0.05, which is 5000 times higher than the value assumed for the calculations of the effect of flashlamps presented in the discussion of light sources in chapter 1.

Azo and triphenylmethane dyes in hydrophobic polymers

Giles, Hojiwala, and Shah (1972) compared the quantum yields for fading of two azo dyes in nylon films and three anthraquinone disperse dyes in acrylic films to the fading of the dyes in solution. The light source used was a mercury arc with three different wavelength regions selected by colored glass filters: 350–500 nm, 440–800 nm, and 500–600 nm.[2] The azo dyes in nylon films had quantum yields for fading of approximately (0.1 to 1) $\times 10^{-5}$ when exposed to 440–800 nm light. These yields were about an order of magnitude lower than the solution yields and are similar to the yield assumed for the example calculations presented in the discussion of light sources in chapter 1. However, one of the anthraquinone dyes—1,4-(dimethylamino)-anthraquinone (C.I. Disperse Blue 14)—had a quantum yield for fading in poly(ethylene terephthalate) acrylic films greater than 10^{-4} for the visible-light (440–800 nm) exposure. This yield is an order of magnitude higher than that shown by the azo and other anthraquinone dyes tested. It suggests that on the order of 10^3 flashes may be the allowable limit for acrylic fibers dyed with this colorant.

As part of a study of oxidation and reduction processes in the fading of dyes by light, Giles and coworkers (1972) also measured the rates of fading as a function of the initial absorbance of several azo and triphenylmethane dyes in nylon, cellophane, and polypropylene films. The most sensitive dye-film combination investigated was C.I. Acid Green 50 in nylon, which was faded approximately 30% after 1.8 hours of exposure to a Microscal mercury lamp in a glass housing. During this time the sample would have been irradiated with approximately 10^{19} photons/cm^2 of near-ultraviolet light, mainly at 365 nm. If a 5% fade is equivalent to the total display lifetime, and flashlamp exposure is to contribute about 0.01 of this, approximately 2×10^{16} near-ultraviolet photons would be the maximum allowable. This total would be reached with no more than 10^4 flashes of a xenon strobe lamp without near-ultraviolet filtering.

Azo and anthraquinone dyes in films of ethyl cellulose, cellulose diacetate, and polyamide

The most sensitive 4-hydroxyazobenzene dye investigated by Tera and coworkers (1985a, 1985b) was the 2′-nitro derivative at the lowest concentration in a polyamide film; the sample faded 10% in 2 hours in a Microscal apparatus with a mercury–tungsten lamp. This corresponds roughly to irradiation by approximately 2×10^{20} visible-light photons/cm^2 and slightly less than 2×10^{19} near-ultraviolet photons/cm^2. Keeping the effect of flash exposure to less than 0.01 of an overall acceptable fading of approximately 2% (for this light shade) would require limiting the total number of flashes to about 4×10^3 from a studio strobe setup without near-ultraviolet filtering.

Vinylsulfonyl derivatives of reactive dyes in cellophane films

Okada, Kato, and coworkers (1990) reported that at very high relative humidity, fading of a reactive vinylsulfonyl azo dye derivative was sensitized by rose bengal or by a reactive phthalocyanine derivative. The sensitization was caused by light from a carbon arc that had been filtered to allow only wavelengths greater than 510 nm to reach the sample. The azo compound does not absorb these longer visible wavelengths. The data suggest that this dye in cellophane films is very susceptible to sensitized fading at a relative humidity approaching 100%. Because of the large number of photons of wavelengths greater than 510 nm emitted by a xenon strobe light, approximately ten flashes might be the maximum allowable for this particular dye combination at very high relative humidities. Fading appears to be less efficient when the relative humidity is lower (Okada, Hirose, et al. 1990b). These latter experiments were performed without using cutoff filters, so a numerical comparison with the earlier results is not possible. For yellow aminopyrazolinyl vinylsulfonyl azo derivative in combination with C.I. Reactive Red 22 (another vinylsulfonyl azo dye), the extent of fading depended on the availability of oxygen in the cellulose films at 100% RH. When wavelengths below 390 nm were excluded, the rate of fading of each dye was about half of what it was when lower wavelengths were included (Okada, Hirose, et al. 1990a). About one-quarter of the absorbance in the top cellophane layer was lost in 4 hours. This corresponds very roughly to an exposure of approximately 10^{18} photons/cm^2 in the 395 nm band of the carbon arc; to approximately 5×10^{17} photons/cm^2 in the 450 nm band; and to slightly less than 10^{18} photons/cm^2 above 500 nm. This level of irradiation suggests that these two dyes, at high relative humidities, could endure thousands of flashes from a xenon studio strobe lamp that has near-ultraviolet filtering. Of the dyes tested, a copper phthalocyanine derivative appeared to be the most sensitive to the longer-wavelength visible light, but complete numerical data were not presented.

Other dyes in film-forming polymers

Zweig and Henderson (1975) investigated the stability of a wide variety of dyes in thin cellulose acetate films that also contained 9,10-dimethylanthracene; the latter compound, which absorbs near-ultraviolet but not visible light, is an efficient trap for singlet oxygen. The films were irradiated with visible light of wavelengths greater than 430 nm from a 500 W (tungsten) quartz–iodine lamp. This slide-projector lamp was 14 cm from the films; however, there was no report of unacceptable heating of the samples. The losses of dye during the exposure time of 1000 seconds ranged from 0% to 89%. Of all the dyes tested (including anthraquinone, azo, methine, xanthine, and phthalocyanine dyes), the most sensitive were the azo dye ethyl red; the xanthine dyes eosin Y and rose bengal; and the methine dye 1,1′-diethyl-2,2′-cyanine iodide. From the data supplied by the authors on the energy of the radiation at the position of the samples, and using the assumption that photoflash exposure should not cause more than 0.01 of a 2% acceptable total lifetime amount of change in these dyes, it can be calculated that no more than

10^3 flashes could be allowed. However, the unusual conditions for these experiments should be borne in mind; museum and archival objects are not commonly "doped" with an efficient singlet oxygen quencher.

McAlpine and Sinclair (1977) investigated the photostability of a series of six different 1,4-disubstituted aminoanthraquinone disperse dyes in nylon, cellulose acetate, or poly(ethylene terephthalate) acrylic films. They showed that the substrate and the relative humidity, as well as the nature of the substituent groups on the dye, can cause major variations in the sensitivity of the dyes to photodecomposition. The most sensitive combination studied, C.I. Disperse Blue 14 in acrylic, was 10% decomposed after 4.5 hours of exposure to a Microscal mercury–tungsten lamp, which includes some near-ultraviolet light. If it is assumed that a 2% change is the maximum acceptable and that flash exposure should contribute no more than 0.01 of this, it can be calculated that approximately 8×10^{16} near-ultraviolet photons/cm^2 and less than 8×10^{17} visible-light photons/cm^2 may be permitted from flash exposure. This is equivalent to about 10^3 flashes if the near ultraviolet, specifically the 365 nm band, is not blocked. The other dye–polymer combinations investigated ranged from 1.5 to over 100 times less sensitive.

Zamotaev, Mityukhin, and Luzgarev (1992) investigated the photoinduced cross-linking and degradation of polyethylene containing a series of halogenated 9,10-disubstituted anthraquinone derivatives or alkyl-anthraquinone derivatives with and without halogen substituents. All the colorants they used had absorption bands in the near-ultraviolet region; this absorption was used to initiate the photochemical reactions. The authors report that the extinction coefficients of the blue absorption bands of these anthraquinones were very low; that is, they were not strong absorbers of blue light. When the colorants were irradiated with ultraviolet light of wavelengths less than 350 nm, radicals with absorption bands in the visible were formed with quantum yields of about 0.25. These radicals subsequently reacted with the polyethylene, causing cross-linking and degradation. Exposure to flash photography when the wavelengths are above 360 nm should not cause photodegradation of these anthraquinone derivatives in polyethylene. However, if these materials are subjected to shorter-wavelength ultraviolet radiation—from a mercury arc or unfiltered photocopier sources, for example—concomitant exposure to blue light could induce deleterious changes in the polymer, as well as causing degradation of the dyes.

The quantum yields for photodegradation of an anthraquinone, a monoazo, and a diazo acid dye in nylon films are reported by Lishan and coworkers (1988). The anthraquinone derivative C.I. Acid Blue 277 had a quantum yield of approximately 10^{-6} when irradiated with 366 nm light from a xenon–mercury-arc lamp. The yield fell to approximately 3×10^{-8} when visible light of wavelengths above 400 nm was used. The azo dyes (C.I. Acid Red 361 and C.I. Acid Orange 156) had slightly lower yields in the visible region. Yields were increased somewhat in the absence of oxygen. The very low yields suggest that these particular dyes in nylon films could withstand on the order of 10^7 flashes from a xenon studio strobe without deleterious effects.

Ball and Nicholls (1984) conducted an extensive investigation of the fading of an azo dye that can exist in either the azo or hydrazone tautomeric form in a substrate—in this case, cellulose acetate and cellulose triacetate films. When 4-phenylazo-1-naphthol was irradiated at 490 nm, only the hydrazone form absorbed the light energy, and permanent fading took place. The azo dye was also faded if methylene blue was included in the films as a sensitizer and if light of wavelength 665 nm was used to excite this sensitizer directly. In contrast, if the azo form of 4-phenylazo-1-naphthol was excited directly, the spectral changes that occurred in the dye were reversible. The authors provide details of the film preparation and dye content and of the xenon light source they used for visible radiation. However, they isolated the specific visible bands with a monochromator and did not indicate how much of the radiated source energy actually passed through the monochromator and reached the samples. If this missing information is approximated, the quantum yield for dye fading may be calculated to have been as high as 0.01 in the absence of sensitizer, and an order of magnitude higher in its presence. These high yields, combined with the numerical data in the report, suggest that up to 10^{10} dye molecules/cm^2 might be faded by each flash from a camera with built-in flash. About 3×10^{11} dye molecules/ cm^2 could be faded if an appropriate sensitizing dye were also present. Reasonable dye concentrations in an object might be approximately 10^{15} to 10^{18} molecules/cm^2 for moderately light to heavy coverage. In the case of lighter coverage, roughly 10^4 camera flashes might cause almost 5% of the dye molecules to fade, using up the entire display lifetime. Gallery flash photography of objects with this particular dye should clearly be restricted. Studio xenon strobe lights may be an order of magnitude more damaging, so studio photography should also be carefully monitored. However, these considerations would apply only to this particular dye, or to a dye with a similarly active hydrazone tautomeric form, in a substrate that permits fluctuations between the two conformations to occur.

A thorough study of the laser-induced photochemical reactions of two triphenylmethane dyes—ethyl violet and crystal violet—bound to the polyelectrolytes poly(methacrylic acid) and poly(acrylic acid) in aqueous solution was conducted by Jones, Oh, and Goswami (1991). They irradiated the samples with pulses of 532 nm light and observed that the extent of fading depended on the ratio of polymer to dye, on the dye concentration, and on the presence of an electron acceptor. When the dye–polymer ratio was low, so that all the dye molecules were surrounded by polymer, the quantum yield in the poly(methacrylic acid) polymer was as high as 0.003. The same type of calculation that was used above for the azo dye was also applied to this case. Several thousand flashes may be the maximum permissible for these oxidizable triphenylmethane dyes when they are present as monomers in charged polymers of this type.

Duxbury (1994a, 1994b) investigated the sensitized fading of acidic and basic triphenylmethane dyes in films of methylcellulose, of poly(vinyl acetate), and of gelatin. Unfortunately, all the work was conducted using ultraviolet light sources that emit more than 90% of their radiation between 280 and 360 nm, so the results are not directly relevant to the present topic. The quantum yields for each step in the photochemical reactions of the dyes—with the exception of C.I. Acid Green 9—are about 3×10^{-5} to 3×10^{-4}. These values are not high, considering that the source included ultraviolet-B light as well as ultraviolet-A. They suggest that these dyes might be fairly stable to moderate photoflash exposure. However, in the absence of specific data obtained with visible-light sources, no firm conclusions can be drawn. The author singles out the behavior of C.I. Acid Green 9 in methylcellulose in the presence of sodium azide, commenting that the yields for photochemical reaction are "very large." Numerical data are not presented for this case. Because museum and archival objects are unlikely to contain the azide reagent, the anomalous behavior of a particular dye in its presence may not be cause for concern.

Matsushima, Mizuno, and Itoh (1995) investigated the photochromism of 4′-amino-substituted 2-hydroxychalcones as model leuco (that is, colorless forms of) compounds for flavylium dyes. Irradiation of solutions or of poly(methacrylic acid) films containing these compounds causes formation of the colored product. Heating of the solutions subsequently induces reversion to the leuco form, but the dye in the polymer films does not completely revert. Data for irradiation with the 405 nm band isolated from a high-pressure mercury arc suggest that it would take approximately 10^6 flashes to cause a detectable change in the dyed materials, so that on the order of 10^4 total flash exposures might be allowed over the display lifetime of an object with these dyes.

Other specific dye–polymer systems

To evaluate the suitability of phenothiazine in a poly(vinyl chloride) film as an ultraviolet dosimeter, Tennent, Townsend, and Davis (1982) tested the response of the dye to gallery illumination or a Microscal lamp. Samples were glazed with a series of commercial plastics that acted as cutoff filters at different wavelengths in the near-ultraviolet region. When the ultraviolet filter Perspex VA was used to block wavelengths below 380 nm (the transmission of this filter rises steeply between 380 and 400 nm, from 0% to approximately 72%), the transmission of the phenothiazine-containing film sample fell linearly with duration of exposure for approximately 90 hours. The sample discolored much more slowly when Perspex VE, a filter that blocks all wavelengths below 400 nm, was used. The decrease in transmission was negligible after 50 hours if a Uvethon Y filter, which blocks wavelengths below 470 nm, was in place. The data for the sample shielded by the VA filter can be used to calculate an approximate allowable exposure of phenothiazine in a poly(vinyl chloride) film to xenon photoflash without complete ultraviolet blocking. This amounts to a few thousand flashes for a studio strobe and more than 10^4 flashes for a point-and-shoot camera.

During a study of electron-transfer reactions in proteinlike matrices, Jones, Feng, and Oh (1995) investigated photoinduced changes in the xanthine dye eosin Y bound to tyrosine and included in poly-(vinylpyrrolidone). Under highly alkaline conditions, the quantum yield for nonradiative decay of excited states of the dye could be as high as

0.1. For example, this behavior was observed at pH 12.0 but not at pH 9.0. The results suggest that significant photodegradation of this dye associated with proteinaceous material may occur only if museum and archival objects are much more alkaline than usual.

Fisher, Lewis, and Madill (1976) investigated the flash photolysis of eosin bound to the water-soluble protein lysozyme. They presented evidence that electron transfer occurs between dye molecules and protein and results in oxidation of the protein as well as changes in the dye. Unfortunately, the light source used in these experiments was a frequency-doubled ruby laser emitting only 347 nm light, so the numerical data cannot be applied to the effects of photoflash.

Seybold and Wagenblast (1989) tested the fastness of derivatives of perylene and violanthrone in poly(methyl methacrylate) polymers to outdoor light. Both of these dye classes have at least five fused six-membered rings. The researchers chose polymers that are used to fabricate fluorescent-light collectors and light concentrators for outdoor applications. With sunlight as the light source, the investigators noted that if wavelengths less than 345 nm were blocked by filters, the dyes underwent only trivial changes even after eleven days of exposure. The observations suggest that these dyes, when included in poly(methyl methacrylate), would probably be stable to extensive photoflash exposures from xenon sources.

Inks on paper

In Lafontaine's experiments (1979), samples of fiber-tipped-pen ink on acid-free paper were exposed to 110–500 hours of light from sunlight fluorescent tubes with and without a plastic ultraviolet filter. After 110 hours of exposure to visible light at 1.7×10^4 lux, the most sensitive inks showed a change of 75–80 units in the color scale used for the experiments. On this scale, a change of 8 ± 2.5 units was the smallest change considered unacceptable by a group of conservators. If a 6-unit change represents the shortest exhibit lifetime of an object with these inks, if flash exposure is to contribute only 0.01 of this amount, and if sunlight fluorescent tubes are equivalent to daylight fluorescent tubes, then on the order of 10^{18} visible-light photons/cm^2 might be the maximum allowable due to flash. This is equivalent to slightly more than 10^3 xenon studio strobe exposures or to 10^5 point-and-shoot camera flashes with the near ultraviolet completely blocked.

Kaneko and Yamada (1981) found that the quantum yield for photoreduction of methyl viologen on paper was approximately 0.001 in air at 460 nm, which is significantly higher than the yield in solution. The authors hypothesize that the nature of the chemical reactions is changed when the dye is adsorbed onto the cellulose. Each flash of a studio xenon photoflash unit might irradiate an object with approximately 3×10^{11} photons/cm^2 at this wavelength. If the yield is the same as that reported in the experiments, approximately 2×10^8 molecules might be reduced. For low concentrations of the dye on paper, for example, 10^{15} molecules/cm^2, a single flash might destroy one dye molecule in 10^7. If destruction of 5% of the dye would consume the total display lifetime of the object, and photoflash is to contribute no more than 0.01 of this, then about 10^4 flashes or several hundred photocopies would be the maximum allowable.

Inorganic pigments associated with synthetic polymeric materials

In the Whitmore and Bailie (1990) study of chalking by resins containing white pigments, the total light dose between 320 and 400 nm required to observe a 10% decrease in the solubility of the resin in the most sensitive case was approximately 100 mW hours/cm^2. This corresponds to roughly 10^{24} photons/cm^2 of near-ultraviolet light. A conservative choice of an acceptable exposure that would cause a negligible change might be 0.001 of this, or 10^{21} photons/cm^2, a dosage approximately 10^6 times higher than the amount of light delivered by flashlamps. This suggests that these pigmented resins could withstand approximately 10^6 flashes from a xenon studio strobe setup without near-ultraviolet blocking.

Fluorescent Whitening Agents

Organic compounds

Using data from Shosenji and coworkers (1983) on the composition of the sample films to calculate the number of molecules per unit area, and using manufacturer's data for the output of a typical high-pressure mercury arc, it is estimated that 10^{15}–10^{16} molecules of methyl violet indicator dye were degraded by approximately 10^{22}–10^{23} photons of 365 nm ultraviolet light that struck the films. This suggests that the quantum yield could be as low as 10^{-7}, which is two orders of magnitude less than that assumed in the example calculations presented in the discussion of xenon light sources in chapter 1. Clearly, more than 10^6 flashes from a xenon strobe lamp could be tolerated by this dye at a similar concentration in the presence of pyrazoline fluorescent whitening agents in the acrylonitrile copolymer films investigated.

In the Oda, Kuramoto, and Kitao (1981) study of dye-sensitized destruction of stilbene fluorescent brighteners, less than 2×10^{20} photons caused a 15% change. If an acceptable loss of optical brightener is on the order of 5%, and exposure to photoflash is to cause no more than 0.01 of this change, then about 6×10^{18} photons delivered by flash exposure may be acceptable. Because the area exposed is not given, the number of photons per square centimeter cannot be determined. A studio xenon strobe delivers approximately 10^{15} visible-light photons/cm^2 in the same wavelength range. These numbers suggest that roughly 6×10^3 would be the maximum allowable number of flashes for an object containing these stilbene optical brighteners and a dye that could sensitize their degradation.

Photographic and Reprographic Materials

Modern photographs

Under the exposure conditions used by Giles and coworkers (1973) to study color transparencies and prints, 5 hours of Microscal exposure was roughly equal to 7×10^{20} visible-light photons/cm^2 and to about 10^{20} near-ultraviolet photons/cm^2. If this exposure caused a just-perceptible change, flash exposure should deliver no more than 0.01 of this amount, or less than 10^{19} visible-light photons/cm^2. From 10^4 to 10^5 flashes of a xenon studio strobe setup would thus be the lifetime limit for prints with this sensitivity, assuming that near-ultraviolet wavelengths are not blocked.

In the study of color transparencies by Townsend and Tennent (1993), a ΔE of 25 was caused by exposure to 3×10^6 lux hours of visible light from a Microscal lamp (the ultraviolet component was filtered out). This was about 25 times a perceptible change. If photoflash is to cause no more than 0.01 of the latter change, the equivalent of only 1200 lux hours of visible light from flashlamps should be allowed. This is calculated to be reached in about 5×10^3 flashes to 10^4 flashes of a studio strobe, if the spectral distribution of the lamp outputs in the visible is assumed to be similar.

Dye photochemistry

From the data of Chatterjee and coworkers (1990), and the fact that a xenon strobe lamp emits approximately 10^{14} photons/cm^2 in a band 40 nm wide centered at 532 nm (the visible wavelength used in the study and absorbed by dyes that photosensitize color film), it is calculated that as many as 10^{13} molecules/cm^2 could be altered per flash. If the coverage of the dye is moderate—for example, 10^{16} molecules/cm^2—and the yield is the same in the film as in solution, then ten flashes would degrade 1% of the dye. Allowing for a possible decrease in yield of 10- to 100-fold in the emulsion would still suggest that a maximum of about one hundred to one thousand flashes could be allowed.

The gelatin-film samples used in the Wilkinson et al. (1993) study contained between 2×10^{-2} and 4×10^{-2} moles of dye per liter in a film that was 10^{-3} cm thick, and the laser beam illuminated an area 0.10 cm in diameter. Thus there were about 10^{14} to 2×10^{14} dye molecules in the volume illuminated by the laser light. Each flash of laser light delivered to this volume about 3×10^{-6} to 7×10^{-6} J of energy as photons of wavelength 585 nm. This is calculated to be equivalent to about 10^{13} to 2×10^{13} photons, using equation 1.3. Because 100 such flashes had chemically altered enough dye molecules that the sample was no longer useful, it is assumed that at least 10% of the dye molecules had been degraded by this exposure. If this is correct, then the yield for photochemical degradation must have been at least 0.005 and could be as high as 0.02. This is calculated as follows: (10% of 2×10^{14} dye molecules) $\div$ (10^{13} photons/flash $\times$ 100 flashes) = 0.02 quantum yield. This is 2000 times the yield assumed in the calculations described in chapter 1 (see "Xenon lamps," pages 15–20). As many as 10^{11} of these dye molecules may be degraded per studio strobe exposure. Depending on the density of the molecules, and assuming that a change in 5% of them would be noticeable, only about 10^3 flashes might be permissible over the lifetime of the film.

Reprography and reprographic materials

Allen, Hughes, and Mahon (1987) observed that the threshold for detection of a significant color change—either as ΔE or as a loss of absorbance—for the crystal violet lactone color-forming system occurred after approximately 1.5 hours of exposure in the Xenotest apparatus. Applying the assumption that flash exposure should deliver no more than 0.01 of the photons that will cause a noticeable change, and using data for a xenon arc in an Atlas Weather-Ometer, it is calculated that approximately 3×10^{17} near-ultraviolet photons/cm^2 are the maximum allowable. This amount would be delivered by roughly 10^4 flashes from a studio strobe that does not have ultraviolet filtering.

The data of Searle (1994) for the red printing ink containing colorant C.I. Pigment Red 53:1 were analyzed as follows: The main visible absorption of the colorant is a broad band between 410 and 550 nm, in which range an Atlas Weather-Ometer xenon arc emits 45% of its visible energy. This is roughly 0.055 W/cm^2. A ΔE of 5, which is the perceptible change that is presumed to indicate the total display lifetime, would occur in about 31 hours if the reciprocity principle is obeyed. Application of equation 1.3 to these data shows that during this time, 1.5×10^{22} photons/cm^2 that could be absorbed by the pigment would have irradiated the object. For flash to provide no more than 0.01 of this amount, or approximately 1.5×10^{20} photons/cm^2, a maximum of about 4×10^5 flashes could be allowed over the display lifetime of the object. However, a photocopier with a 1980s-style xenon light source would expose a print to about 2×10^{17} visible-light photons/cm^2 and to about 3×10^{15} near-ultraviolet photons/cm^2 for each copy made. Thus, a total of only about 700 photocopies would be the maximum allowable over the display lifetime of a print with this ink.

In her investigation of color photocopies and faxes, Norville-Day (1994) observed a ΔE of greater than 16 in the brown-colored sample after it was exposed to 1.6×10^4 lux of Philips 94 fluorescent light for 504 hours. The spectral energy distribution published by Thomson (1986) permits calculation of the total dose as approximately 4×10^{22} photons/cm^2 in the visible region. The change is about eight times that which would be noticeable. Again, applying the 0.01 factor for photoflash exposure suggests that a total allowable exposure would be 5×10^{19} photons/cm^2. This maximum would be delivered by approximately 1.5×10^5 flashes from a xenon strobe setup. The black-covered areas of faxes made with the same type of machine that Hockney used for his faxes suffered a ΔE of almost 38 from the same fluorescent-lamp exposure, suggesting that these documents might be restricted to a total of approximately 5×10^4 flashes from xenon strobe lamps during their display lifetimes.

In his discussion of irradiation conditions during laser digitization, Messier (1997) states that a helium–neon laser (633 nm wavelength) delivering 5×10^{-3} W of power to loci spaced 0.5 mm apart would scan a surface at a typical rate of 0.25 second per location. In that time, the laser would deliver 4×10^{15} photons of red light, according to equation 1.3. If there is no overlap of irradiated areas during the scan, this number of photons falls on each scanned location, which has an area equal to at most 0.05 cm^2. Thus, the irradiated surface receives at least 4×10^{15} photons per 0.0025 cm^2 of area , or 1.6×10^{18} photons/cm^2/scan. Colorant surface concentrations may typically range from 10^{15} to 10^{19} molecules/cm^2. If yields for photochemical change are in the range of 10^{-5} to 10^{-3}, then from one to one hundred scans could consume a lifetime of exposure to flash sources in the regions where highly light-sensitive blue colorants are present at low coverage.

Miscellaneous Objects and Combinations of Materials

Ethnographic objects

In Horie's 1990 study of budgerigar feathers, the samples were exposed to a Microscal mercury–tungsten lamp with and without filtration of the ultraviolet component. Only the gray-green feather showed simple, progressive bleaching in light. Several hundred hours were required to cause a ΔE of 2, with or without near-ultraviolet light blocked. The responses of the other feathers were complex, with ultraviolet light having a significant effect on the bright colors. Exposure of these feathers to visible light for 250 hours caused unacceptable changes in the reflectance spectra. During this time, the Microscal lamp would have delivered on the order of 10^{22} visible-light photons/cm^2. If flash exposure is to be kept under 0.01 of this amount, then 10^{20} photons/cm^2 would be the maximum. This corresponds to allowing about 10^6 flashes from a point-and-shoot camera with ultraviolet blocking, or to about 10^5 studio strobe exposures. Uncertainty regarding the role of near-ultraviolet wavelengths could make this conclusion inappropriately lenient for sources that do not have near-ultraviolet blocking.

Notes

1. The authors do not specifically identify the lamp, but RF5000 series instruments such as theirs had this source.
2. The researchers note as an interesting aside that they used all-glass, broadband filters rather than narrow band-pass filters based on dyed gelatin because the latter would have faded from extensive exposure to the light source.

Chapter 4
Discussion

The conservation literature contains few reports of experiments designed specifically to address the effects of photoflash and reprographic light on art and archival objects. One of the first papers was by Hanlan (1970), but the art conservation community had to wait nearly two and a half decades before Neevel (1994) and then Saunders (1995) added significant contributions to the field. Using different approaches, they each emphasized the importance of comparing both the spectral distribution of the source and the flash dose to gallery lighting. In addition, Saunders tested colorants known to be highly light sensitive. Reports of other investigations that specifically compare the effects of photoflash or reprographic light exposure to display lighting exposure were not located. Also absent from the literature are evaluations of the possible deleterious effects of photoflash and reprographic light exposure as an integral component of risk assessment for collections.

The basic science literature was consulted for technical information on the photochemistry of materials in museum and archival collections. Results of flash photolysis and other photochemical studies can be related to the effects of photoflash, provided that the light sources used are not powerful enough to cause two-photon phenomena; that the spectral distribution is appropriate; and that the experimental sample conditions are similar or can be extrapolated to those of materials in collections. The latter means, for example, that results of experiments performed on solid materials in air, or on colorants on fibrous substrates or in films, should provide relevant information. Estimations based on data from some experiments of these types are included in this review. However, most of the published information used to evaluate the possible effects of photoflash exposure was obtained from lightfastness studies. In many of these investigations, the spectral output of the light source was not precisely defined, or near-ultraviolet wavelengths were present in a different proportion—or to a greater extent—than in the photoflash sources, or both. This situation required the use of assumptions and approximations to obtain the numerical predictions of allowable flash exposures provided in chapters 2 and 3; it meant that in most cases only order-of-magnitude suggestions could be given as guidelines.

Despite these limitations, a large amount of useful information emerged from this literature search and evaluation. When applied within appropriate contexts, these data offer some guidance to museum professionals who must judge the relative magnitude of risk to art and archival materials of flash photography and reprography performed with various light sources.

Specific indications that some restriction should be considered for exposure to photoflash and reprographic light include the following situations: (1) a significant amount of ultraviolet radiation is emitted by the source; (2) materials in the object do not obey the reciprocity principle; (3) materials in the object have a very low lightfastness rating on the Blue Wool scale; (4) materials are suspected of exemplifying the early use of a particular class of product or technology; and (5) objects contain materials similar to ones known to be very light sensitive.

The range of possible responses that art and archival objects can show on exposure to these intermittent light sources is huge. The majority of substances that might be encountered in collections are not likely to be significantly affected by moderate, or what would be considered normal, exposure. This conclusion was reached even though the literature search did not include materials expected to be the most stable to light, such as stone, metal, or ceramic. It remains valid even when a source includes some near-ultraviolet light of wavelengths at or above 360 nm. In contrast, a few materials were found to be sufficiently sensitive that they should not be exposed to any unnecessary light from photoflash or reprographic sources.

Wide ranges of possible responses to photoflash and reprographic light are found within most material categories, and these responses depend on the properties of the individual substances. Figure 4.1 presents

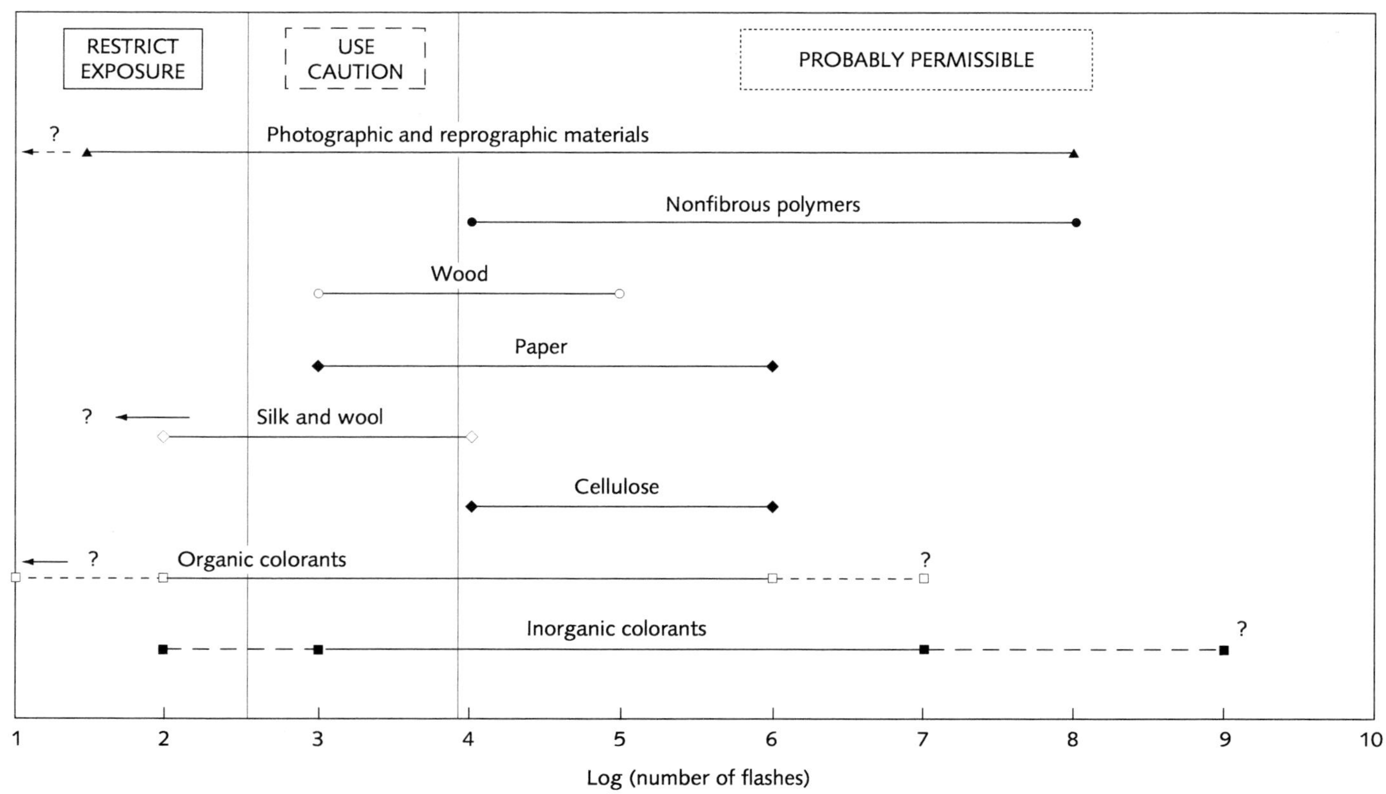

Figure 4.1
Ranges of approximate numbers of allowable flashes from a xenon camera flashlamp (including near-UV wavelengths above 360 nm) that would consume about 0.01 of the total display lifetime for different types of materials. The abscissa is a logarithmic scale (that is, the power of 10 of the number of flashes is plotted); each unit on this scale represents ten times as many flashes as the next lower unit. The divisions between risk categories are intended as suggestions only. Allowable numbers of copies made on a photocopier should be lower by a factor determined by the ratio of the copier source intensity to the photoflash lamp intensity (typically ×10 or greater).

a summary of the order-of-magnitude estimates of the numbers of flash exposures considered safe for the different classes of materials surveyed. As detailed in chapters 2 and 3, most of the numbers were calculated using estimated values for some parameters in conjunction with published numerical data. The results are conservative estimates, as indicated by the nature of some of the assumptions described in the paragraphs that follow. To enhance the usefulness of these figures to those who must evaluate the risks entailed in exposure of sensitive objects to photoflash, the steps taken to evaluate the data are reviewed here.

Assumptions Used for the Calculations

The effect of exposure to photoflash or reprographic light should be evaluated within the context of the total display lifetime of the individual object. The shorter the display lifetime of the object, the more likely that limitations on light exposure are appropriate. In any particular case, the possibility that deleterious effects may result from exposure to photoflash or reprographic light will depend on the chemical and physical properties of the exposed materials and on the spectral distribution, intensity, and duration of the light compared to display lighting.

The existence of a finite display lifetime of art and archival materials is determined by the fact that a fraction of the absorbed photons will lead to a chemical change in molecules of the material. This process will eventually affect enough molecules to cause an unacceptable change in the appearance or integrity (or both) of the object. For many materials, the probability of photochemical change is exceedingly small, and the display lifetime is correspondingly long. For other materials, the probability of photoinduced chemical reaction is relatively high. For all materials, however, the more light that is absorbed, the more quickly the object will reach its lifetime limit. The display lifetime of an object is defined by a perceptible alteration in appearance or an unacceptable change in mechanical stability. It has been assumed for this project that exposure to flash and reprographic light should consume no more than 1% of the display lifetime. These limits were applied directly in the calculations in chapters 2 and 3. In other cases, a numerical definition of the total display lifetime had to be assumed, based on effects of light exposures of known length. Exposures that caused a ΔE of 1 for white and pale colors or a ΔE of 5 for bright colors and deeper shades were used. Alternatively, an exposure that induced a just-perceptible fade in the relevant Blue Wool standard was applied if this comparison was the only numerical information reported. The total acceptable flash exposure over the entire display lifetime was then chosen to be 0.01 of this latter value.

The fastness behavior of Blue Wool standards is frequently used as an experimental control against which the lightfastness of objects containing colorants can be compared. Although the Blue Wool standards have been designed for monitoring exposure to daylight, with the implied inclusion of ultraviolet light, the widely quoted total exposures that cause a just-perceptible fade for these standards are given in photometric units. That is, the exposure ratings suggest numerical values for only the visible light from the source (see, for example, Colby 1992) and do not take into account the ultraviolet light emitted. In contrast, most of the light sources encountered in the literature review emitted significant (but differing) amounts of energy in the near-ultraviolet region. Thus, the resulting flash exposures calculated on the basis of actual fading data for a Blue Wool standard might be more appropriate for sources that include near-ultraviolet light than for those with ultraviolet completely blocked. Conversely, if Blue Wool fade data obtained with a light source lacking ultraviolet is subsequently used to predict fading induced by a light source that does emit ultraviolet as well as visible light, the predicted allowable exposure might be an overestimate.

The results of the studies of fading of Blue Wool standards that are summarized in chapter 1 appear to be inconsistent. The variations in fading behavior may possibly be due to the variations in sensitivity that the standards display toward different wavelengths of light in the visible and near-ultraviolet regions, as suggested earlier by McLaren (1956). Additional experimental work is needed to thoroughly clarify the wavelength dependence of the light sensitivity of the Blue Wool standards.

Other assumptions used repeatedly in the evaluations were the approximate area illuminated by a flash from a xenon studio strobe lamp and, in some cases, the location of the short-wavelength cutoff of near-ultraviolet light. The latter assumption can introduce a fairly large uncertainty because many manufacturers of handheld cameras are unable or unwilling to specify the wavelength cutoff for the built-in flash in their products. An additional problem with this particular assumption is the fact that much of the lightfastness research was conducted with light sources that include ultraviolet-A wavelengths between 320 and 360 nm, as well as from 360 to 400 nm. Many of the materials considered in this report are more sensitive to these wavelengths than they are to visible light. They would thus be able to withstand more flashes than calculated in chapters 2 and 3 *if filters are in place to block the ultraviolet light emitted by the xenon flashlamp.*

Because of the need for these assumptions, most calculations reported in this survey provide no greater accuracy than order-of-magnitude values. Using the data in a more precise way would be inappropriate. One cautionary note, illustrated by some examples in chapters 2 and 3, is that otherwise similar materials can have very different lightfastnesses, so an extension of one particular result to another similar compound may not always be appropriate. Substantial variations within a class of materials can be expected.

Assessing Risk

If the absorption spectra, the quantum yields for photochemical change, and the density of light-sensitive molecules in an object are known, then calculations like those made here can be performed to determine the risk to that object from exposure to any well-characterized light source. In actual situations, however, the time required to identify the materials and obtain the necessary data is likely to be an unattainable luxury. Real-life risk assessment for individual objects will usually require more rapid decision making with limited information available. A useful, approximate rule of thumb may be that the shorter the allowed display time and the lower the display illumination advised on the basis of experience, the more limited the exposure to flash and reprographic light sources should be. This guideline could become significantly distorted if the proposed photoflash light source has a very different spectral output than the illumination source on which the display experience is based.

When faced with the need to state a numerical limitation on flash exposure, a museum professional might use the following approach to evaluate a specific situation. The important first step is to determine if the spectral output of the proposed flash or reprographic light source differs from the display lighting. Does it emit substantially more near-ultraviolet light? Is the shape of the visible spectral output curve of the flash or reprographic light source very different from that of the display light source? (An example might be strong near-ultraviolet and blue bands emitted by a fluorescent tube compared to incandescent gallery lighting.) If so, are these differing wavelengths absorbed by the materials in the object, and do they lead to photochemical alterations in these materials?

If the answers to these questions are yes, then a calculation to determine the photons per unit area reaching the object per flash can be attempted. If the result is more than a few percent of the number of absorbing molecules per unit area of the object, restriction of flash exposure may be appropriate. If there is insufficient information for an approximate calculation, or if the identity of the material is not known but is likely to belong to a class of substances that includes materials with a notably high light sensitivity, caution is suggested. With this information in hand, the museum professional can discuss with the exhibition curator, who may become involved in the final decision, the feasibility of allowing flash photography of a particular object. The importance or necessity of allowing flash photography can then be weighed against the resulting loss of some display time for the object.

If the photoflash light source does not differ qualitatively from the display lighting but is more powerful, the effect of flash exposure is easily determined. The light energy per unit area reaching the object per flash is compared directly to the energy per unit area from the display lighting, making sure the units are the same. Ideally, the total flash exposure (energy per flash multiplied by the number of flashes) in a given amount of time (for example, one day of exhibit) should contribute no more than 0.01 of the display exposure during that same period. Then the number of allowable flashes during that time is calculated as follows:

$$\text{no. of flashes} = \frac{0.01 \times (\text{display energy})}{\text{energy per flash}} \qquad (4.1)$$

The conservator and collections manager or curator must then decide if unrestricted gallery photography will exceed this limit, based on the projected popularity of the object. In the case of a highly sensitive material, for which gallery photography is forbidden, a similar calculation should be made for studio photography, and measures taken to reduce exposures to the minimum required to obtain necessary photographs of the object.

A few extremely light-sensitive materials were identified during the literature review (see fig. 4.1). No substances were found to be so sensitive to visible light, however, that only one or a few flashes from a xenon photographic flashlamp with ultraviolet blocking below 360 nm would cause an unacceptable change in the object. Most highly sensitive materials are likely to be one of two types: (1) early photographs made with experimental processes, such as William Henry Fox Talbot's halide-fixed images, or modern experimentally processed prints that have not been treated for adequate removal of developer and fixer chemicals; and (2) objects incorporating certain natural dyes, or early synthetic dyes that were marketed while still experimental and before fastness testing was routine. If any of these objects have a history of extended light exposure—from extensive display, for example—they have probably already faded significantly. When studio photography is required of an object containing such highly sensitive materials, especially if it is in pristine condition, use of a stand-in during setup is advisable so the original receives the minimum exposure possible. The sensitivity of such objects dictates a highly restrictive policy with respect to all nonessential light exposures, including gallery photography and reprographic processes. A record of all light exposures, including those in the conservation and photography studios, should be maintained.

A new approach to determining how sensitive objects are to light exposure is being developed by Whitmore and his colleagues (Whitmore, Pan, and Bailie, 1999). These researchers have constructed a microfading apparatus that uses a xenon light source with ultraviolet-blocking filters and focusing lenses to expose a microscopic area of an object to intense light. The instrument simultaneously monitors the reflection spectrum of the exposed area as it is irradiated. The amount of light energy required to cause a given amount of appearance change is measured. This result could be converted to number of flashes from a xenon photoflash source. The area of the object exposed to light is too small to be noticeable by eye. Other possible long-term effects of this irreversible test have not yet been documented.

The rate at which materials are altered by light may change as exposure time accumulates (see Giles and Forrester 1980). For many substances, the rate of change *decreases* after an initial period during which a significant change in appearance occurs. In these cases, an object that has already been exposed extensively to light will not be affected to the same extent by additional exposure, and flash photography would be less likely to cause a further, unacceptable change. An example of such objects is Japanese woodblock prints containing fugitive organic colorants. In contrast to this type of behavior, a few substances have been identified that show an *increased* rate of fading with time of exposure. Examples of these materials are some azo dyes in polymers (Giles, Johari, and Shah 1968; Allen 1989); a colored, transparent glaze (Whitmore and Bailie 1997); and some pigments in printing inks (Schmelzer 1984).

These materials do not follow the reciprocity principle. The importance of knowing the exposure history of individual objects containing such colorants or formulations cannot be overemphasized. If the anticipated flash exposure will include wavelengths that are absorbed, then unsupervised photography probably should be restricted, especially if the exposure history of the object is not known.

Tremendously popular objects that serve as cultural icons or primary sources may be put at relatively high risk by their unusual popularity, even if their component materials are moderately stable to photoflash and reprographic light. Unrestricted photoreproduction of these materials could easily result in thousands of photoflash (or dozens of photocopy) exposures per day. As a result, unacceptable change would occur much sooner than expected on the basis of typical exposure rates for less popular items. This situation can require an unpopular decision on the part of the conservator, who must act as an advocate for the object rather than for the public in such cases.

Some ultraviolet light sources used for laboratory examination of objects, as well as photocopier light sources, have the potential to be much more damaging than xenon photographic flashlamps. Not only are they likely to radiate more energy per exposure than camera flashlamps, but they also irradiate the object with significant amounts of ultraviolet wavelengths at and below 360 nm. Examples are given in chapter 2 of some colorants and certain papers that could sustain significant damage from modest exposures to these sources. Lakes prepared from fugitive organic yellow and orange colorants, in particular, may be very sensitive to the strong 365, 405, and 435 nm bands emitted by mercury arcs used in ultraviolet photography. It might be advisable not to subject objects with these pigments to this procedure at all, unless the information it provides is essential to art history scholarship or to decisions regarding conservation treatment. Staff personnel responsible for the management of archival collections might find it useful to formulate a photocopying and scanning policy that incorporates assessment of risk from light exposure for the more sensitive materials in their care.

Suggestions for Further Action

Results of the literature search, and the evaluations provided here, indicate that additional experimental investigations of three types would provide information particularly useful for assessing risk due to photoflash and reprographic light exposures. First, researchers typically publish photochemical and lightfastness data from studies on new samples in good condition. Conservators, in contrast, routinely deal with objects that are partly degraded or otherwise altered from their original state; thus, they need more information on the response to photoflash exposures of materials in these conditions. Second, the effects of photoflash and reprographic light exposures on some materials, and on combinations of materials, need to be more clearly defined by experimental studies, particularly for the varied conditions of art and archival objects and for ambient museum environments. Third, information on the spectral output of light sources and on the spectral characteristics of filtering materials, as well as improved means of standardizing the measurement of these quantities and monitoring light dosage, must be readily available to museum staff who need these data for the risk assessment process.

Investigation of the effects of light exposure on already degraded materials in art and archival objects should be a major ongoing activity of conservation scientists. Examples of topics to include in this effort are determining the lifetime of repeatability for photochromic cycling in, for example, blueprints and lignin-containing papers; evaluating materials for which the rate of photodegradation increases with increasing light exposure; determining the effects of photoflash and reprographic light exposure on photographs and ligneous papers that already show some deterioration due to inherent sensitivity or nonoptimal processing; and evaluating the photodegradation of aged historic colorants that are available for sampling.

The photochromic effect—the cycling of fading in light and return of color in the dark—of blueprints and cyanotypes is widely recognized. However, no experimental evidence was located during the literature search regarding the lifetime of the ability of these images to regain color in the dark. Research to determine the extent of reversibility of fading should be undertaken, on both historic and modern materials. The qualitative similarity of any accelerated cycling treatment used to simulate the photochromic effect in the experiments to the actual chemistry that occurs in displayed and stored objects must be ascertained so that the experimental results are valid.

The extent of the photochromic effect observed in some ligneous papers needs to be quantitated with respect to the relative amounts of near-ultraviolet and visible light in display lighting. The short- and long-term effects of display after papers have been exposed to photocopier and scanner light sources with their higher near-ultraviolet content should be investigated. These studies would provide data relevant to maximum allowable photocopy exposures for papers containing lignin. They would also better characterize the photochromic behavior of ligneous papers and the contribution of the photochromic effect to the eventual degradation of such materials.

If consumable historic samples made from early versions of technologies—especially product combinations, such as colorants in textiles and other polymers—become available, researchers should take advantage of this unusual opportunity. Samples of the materials could be used to characterize the effects of gallery lighting with and without exposure to such sources as studio flashlamps, photocopiers, or digital imaging devices. If enough sample is available, pre-aging under simulated bad storage or lighting conditions might be included in these experiments.

The responses of materials that are suspected of not obeying the reciprocity principle when exposed to display lighting combined with flash and reprographic light sources require further study. The materials of concern are those for which the rate of degradation increases with time or goes up suddenly after a period of light exposure. Some examples of these substances, which are mentioned in chapter 2, include lead(II) oxide; some photographs on resin-coated paper from the 1970s; clear glazes and printing inks containing certain colorants such as C.I. Pigment Red 53:1; some impure vermilion preparations that have begun to discolor; halide-fixed photographic images; incompletely or improperly developed and fixed photographs; and certain color films. The work of Whitmore and Bailie (1997) serves as an example of a suitable approach to experimental design; the introduction of intermittent exposure to reprographic light sources could be added to their protocol.

Anecdotal evidence indicates that dyed and weighted silks are extremely fugitive to light; they may be even more sensitive than the research results discussed in chapter 2 suggest. Further investigation of this topic would be appropriate, using well-characterized historic and modern samples. The occasional use of photocopying to record patterns in embroidered silks and in laces made of silk should be strongly discouraged.

Research on the photosensitized degradation of one colorant by another in mixtures of two or more dyes on textiles, in lakes, and in modern polymeric materials should be continued. Mixtures favored by artists should be emphasized. The advice of experienced color chemists might be sought in choosing historic and contemporary colorant combinations that are likely to lead to photosensitization effects.

The extreme light sensitivity of some colorants, such as the natural dyes turmeric and orchil and some magenta colorants used in color photographs, is widely recognized. Conservators already take significant precautions to restrict unnecessary light exposure of objects containing these colorants, including forbidding gallery photography and limiting studio photography when possible. Studies designed to further characterize the fugitive nature of these materials are unlikely to add new *qualitative* information to the general understanding of the behavior of these substances, and thus collection of additional data is of a lower priority than other research topics suggested here.

Investigation of the degradation of various types of photographic materials is currently an active field (see, for example, Zinn et al. 1994). However, more emphasis on the effects of light from photocopiers and light sources in other types of imaging equipment would be welcomed. Photographic materials that have been improperly processed or already subjected extensively to display lighting (or both) need to be included in this work.

Characterization of the light sensitivity of contemporary dyes used in color photographic processes, and of the colorants in inks used in modern printing processes such as ink-jet printing, should be an ongoing effort. A few of these substances have demonstrated marked sensitivity to visible light, and their fastness to near-ultraviolet light is suspect. They should be tested in photographic and print formulations that mimic their actual condition in objects found in collections. Not only does the lightfastness of these materials require documentation, but their response to reprographic light before or after exposure to display lighting also needs to be studied. An example of appropriate samples for investigating the light sensitivities of contemporary color-printing process inks is the set of prints, with their color separation strips, published by Lowe and coworkers (1997). This artist and his colleagues have reproduced the same image using several different techniques, all to the same exacting standards and in as uniform a manner as possible.

The response to light of modern materials that are not of archival quality but are used by artists should also be characterized. Examples are Iris ink-jet colorants, colored pencils, marker pens, and some photographic materials and processes. Unexpected combinations of materials should also be considered in this category.

Research should be undertaken to document how fluorescent whitening agents, such as titanium dioxide and organic optical brighteners, can act as catalysts to photosensitize the fading of colorants and the degradation of paper and textile supports. Pre-aged as well as fresh samples should be tested for their response to continuous light and to flash and reprographic light sources. It is particularly important to investigate whether the response of objects containing fluorescent whitening agents to near-ultraviolet light changes abruptly when the optical whitener is consumed.

The effects of photoflash and reprographic light on materials that have already been exposed not only to light but also to atmospheric pollutants should be studied. Larsen, Vest, and Kejser (1994) have called for this type of research to be performed on dyed leathers. Investigations should be extended to other materials that are found in art and archival objects and are likely to be sensitive to pollutants.

Research by Neevel (1994) on the irradiances of flashguns and photocopier light sources need to be extended to light sources used in contemporary photocopiers, scanners, and digital imaging procedures and to light sources used for flash photography in the United States. This work should be ongoing; new sources should be characterized as they come into use. A database with this information might be maintained and made available to conservators. The transmission characteristics of new glazing products and of near-ultraviolet and color-correcting filters for flashlamps should also be measured and maintained in this manner.

The fading behavior of the Blue Wool standards requires further clarification if these indicators are to serve as truly reliable dosimeters in different lighting conditions. The work of Bullock and Saunders (1999) should be extended to include more detailed spectral characterization of the light sources used to induce fading. Both action spectra and kinetics of fading caused by light of different wavelengths in the visible and near-ultraviolet regions should be measured. Ideally, this project would be undertaken by a laboratory equipped with state-of-the-art light-monitoring devices as well as the types of light monitors more commonly available to conservators, so that correlation between precise numerical data and measurements actually made in galleries can be emphasized.

The recent interest in the development of simple, reproducible dosimeters for monitoring light exposures that art and archival materials receive in galleries should be strongly encouraged. The approaches of Lavédrine, Gillet, and Garnier (1999), to devise a sturdy indicator that changes color after absorbing relatively low doses of visible and near-ultraviolet light; and of Bacci and coworkers (1999), to construct dosimeters from fugitive materials found in particular collections, can each be very useful. More research is needed to ensure that the ultimate products perform reliably in various conditions without undue sensitivity to changes in environmental parameters they are not designed to measure. Use of indicators such as these (or of thoroughly characterized Blue Wool standards) could then become routine practice in conservation studios as well as in exhibition spaces.

With experience, conservators and collections managers become highly knowledgeable about the general vulnerability to light of the objects in their care. However, these museum professionals do not have the time to obtain the precise photochemical and spectral data required to make photoflash and reprographic exposure decisions on a case-by-case basis. They cannot be expected to keep up with all the new developments in understanding the photochemistry of the myriad substances now found in art and archival collections. As this survey shows, the published technical literature rarely contains data that can be used directly to state maximum allowable flash exposures. To fill this need, it is recommended that literature reviews of this topic be conducted periodically and the results published. The form of publication envisioned is that of a review article that includes not only abstracts of papers but also brief evaluations of the data.

This publication has initiated that literature review process. The results of future evaluations should be made as widely available as possible to members of the conservation community. In this manner, collections care professionals responsible for risk assessment will have timely information available regarding exposure of art and archival objects to strong intermittent light such as that from photoflash and reprographic sources.

References

Aaland, M., and R. Burger
1992. *Digital Photography*. New York: Random House.

Albini, A., E. Fasani, and S. Pietra
1982. The photochemistry of azo-dyes: The wavelength-dependent photoreduction of 4-diethylamino-4′-nitroazobenzene. *Journal of the Chemical Society, Perkin Transactions* 2: *Physical Organic Chemistry* 11: 1393–95.

Allen, N. S.
1987. Photofading mechanisms of dyes in solution and polymer media. *Reviews of Progress in Coloration* 17: 61–71.

1989. Effects of dyes and pigments. Chap. 20 in *Comprehensive Polymer Science*, ed. G. Allen and J. C. Bevington. Vol. 6, sec. 20.1–20.4. Oxford: Pergamon Press.

1996. Photoinitiators for UV and visible curing of coatings: Mechanisms and properties. *Journal of Photochemistry and Photobiology A: Chemistry* 100: 101–7.

Allen, N. S., J. P. Binkley, B. J. Parsons, G. O. Phillips, and N. H. Tennent
1983. Photofading of anthraquinone disperse dyes in epoxy resin films: Photoreduction, energy transfer and photostabilisation processes. *Dyes and Pigments* 4: 11–24.

Allen, N. S., and M. Edge
1992a. Biodegradable and photodegradable polymers. Chap. 6 in *Fundamentals of Polymer Degradation and Stabilisation*, 131–48. London: Elsevier Applied Science.

1992b. Effects of dyes and pigments. Chap. 7 in *Fundamentals of Polymer Degradation and Stabilisation*, 149–75. London: Elsevier Applied Science.

1992c. Lightfastness of dyes and pigments. Chap. 8 in *Fundamentals of Polymer Degradation and Stabilisation*, 176–93. London: Elsevier Applied Science.

1992d. Photodegradation and photooxidation of polymers. Chap. 4 in *Fundamentals of Polymer Degradation and Stabilisation*, 75–101. London: Elsevier Applied Science.

Allen, N. S., K. O. Fatinikun, A. K. Davies, B. J. Parsons, and G. O. Phillips
1981. Factors influencing the photofading of commercial anthraquinone dyes in solution. *Dyes and Pigments* 2: 219–29.

Allen, N. S., J.-L. Gardette, and J. Lemaire
1982. Interaction between titanium dioxide pigments and hindered piperidine/antioxidant combinations in the photostabilisation of polypropylene. *Dyes and Pigments* 3: 295–305.

Allen, N. S., N. Hughes, and F. Mahon
1987. Photochemical fading and photostabilisation of the crystal violet lactone colour former system. *Journal of Photochemistry* 37: 379–90.

Allen, N. S., and J. F. McKellar
1978. Photochemical reactions in commercial poly(ethylene 2,6-naphthalate). *Journal of Applied Polymer Science* 22: 2085–92.

1980a. Photosensitized degradation of polymers by dyes and pigments. Chap. 5 in *Photochemistry of Dyed and Pigmented Polymers*, ed. N. S. Allen and J. F. McKellar, 247–78. London: Applied Science Publishers.

1980b. Structural influences on the photostability of anthraquinone dyes in man-made polymers. Chap. 7 in *Developments in Polymer Photochemistry: I*, ed. N. S. Allen, 191–215. London: Applied Science Publishers.

Allen, N. S., J. F. McKellar, and B. Mohajerani
1980. Lightfastness and spectroscopic properties of basic triphenylmethane dyes: Effect of the substrate. *Dyes and Pigments* 1: 49–57.

Allen, N. S., J. F. McKellar, and S. Protopapas
1977. Lightfastness, phototendering and fluorescence of 3-methoxybenzanthrone. *Journal of Applied Chemistry and Biotechnology* 27: 269–73.

Allen, N. S., B. Mohajerani, and J. T. Richards
1981. Photofading of basic triphenylmethane dyes: Evidence for electron transfer. *Dyes and Pigments* 2: 31–35.

American Association of Textile Chemists and Colorists (AATCC), Midwest Section
1990. Influence of fluorescent whitening agents (FWA) on silk phototendering and dye fading. *Textile Chemist and Colorist* 22(5): 17–21.

Andrady, A. L.
1993. Polymer Materials. Chap. 7 in *UV-B Radiation and Ozone Depletion: Effects on Humans, Animals, Plants, Microorganisms and Materials*, ed. M. Tevini, 193–227. Boca Raton, Fla.: Lewis Publishers.

1997. Wavelength sensitivity in polymer photodegradation. *Advances in Polymer Science* 128: 47–94.

Andrady, A. L., K. Fueki, and A. Torikai
1990. Photodegradation of rigid PVC formulations III: Wavelength sensitivity of the photo-yellowing reaction in processed PVC formulations. *Journal of Applied Polymer Science* 39: 763–66.

Andrady, A. L., and N. D. Searle
1995. Photoyellowing of mechanical pulps. Part 2: Activation spectra for light-induced yellowing of newsprint paper by polychromatic radiation. *Tappi Journal* 78: 131–38.

Andrady, A. L., N. D. Searle, and L. F. E. Crewdson
1992. Wavelength sensitivity of unstabilized and UV stabilized polycarbonate to solar simulated radiation. *Polymer Degradation and Stability* 35: 235–47.

Andrady, A. L., Y. Song, V. R. Parthasarathy, K. Fueki, and A. Torikai
1991. Photoyellowing of mechanical pulp. Part 1: Examining the wavelength sensitivity of light-induced yellowing using monochromatic radiation. *Tappi Journal* 74: 162–68.

Andrady, A. L., A. Torikai, and K. Fueki
1989. Photodegradation of rigid PVC formulations I: Wavelength sensitivity to light-induced yellowing by monochromatic light. *Journal of Applied Polymer Science* 37: 935–46.

Anton, A.
1965. Determination of hydroperoxides in ultraviolet-irradiated Nylon 66. *Journal of Applied Polymer Science* 9: 1631–39.

Atlas Electric Devices Co.
1996. Xenon Weather-Ometer Ci 4000 Instruction Manual.

Bacci, M., M. Picollo, S. Porcinai, and B. Radicati
1999. Indoor environmental monitoring of colour changes of tempera-painted dosimeters. In *12th Triennial Meeting, Lyon, 29 August–3 September 1999: ICOM Committee for Conservation Preprints.* Vol. 1, 3–7. London: James & James.

Baer, N. S., A. Joel, R. L. Feller, and N. Indictor
1986. Indian yellow. In *Artists' Pigments: A Handbook of Their History and Characterization*, ed. R. L. Feller. Vol. 1, 17–36. Washington, D.C.: National Gallery of Art.

Bagnall, R. S., ed.
1995. *Digital Imaging of Papyri: A report to the Commission on Preservation and Access.* Washington, D.C.: Commission on Preservation and Access.

Ball, P., and C. H. Nicholls
1984. The role of azo–hydrazone tautomerism in the photofading of 1-phenylazo-4-naphthol in polymer substrates. *Dyes and Pigments* 5: 437–55.

Banik, G., and J. Ponahlo
1983. IRS-Untersuchung der degradierenden Wirkung von Licht und SO_2 auf pigmentiertes Papier (Infrared investigation of the degrading effect of light and SO_2 on pigmented paper). *Mikrochimica Acta* (Wien) 1: 87–94.

Baumgarte, U., and D. Wegerle
1986a. Veränderungen von Cellulosefärbungen durch Lichteinwirkung (Changes in cellulose dyes due to the influence of light), sec. 1–3.1. *Melliand Textilberichte* 67(8): 567–572.

1986b. Veränderungen von Cellulosefärbungen durch Lichteinwirkung (Changes in cellulose dyes due to the influence of light), sec. 3.2–3.3. *Melliand Textilberichte* 67(9): 640–647.

Becker, M., and N. Tuross
1992. Novel approaches for assessing the preservation of historic silks: A case study of the first ladies' gowns. *MRS Bulletin* 17: 39–44.

1994. Initial degradative changes found in *Bombyx mori* silk fibroin. Chap. 22 in *Silk Polymers*, ed. D. L. Kaplan, W. W. Adams, B. Farmer, and C. Viney, 252–69. Advances in Chemistry Series, no. 544. Washington, D.C.: American Chemical Society.

Bergthaller, P.
1985. Light stability improvements in dye diffusion photography by use of metallizable azo dyes. In *Second International Symposium: The Stability and Preservation of Photographic Images, the Public Archives of Canada and the Delta Ottawa Hotel, Ottawa,Ontario, Canada, 1985,* 47–50. Springfield, Va.: Society of Photographic Scientists and Engineers.

Berrie, B. H.
1997. Prussian blue. In *Artists' Pigments: A Handbook of Their History and Characterization,* ed. E. W. Fizhugh. Vol. 3, 191–217. Washington, D.C.: National Gallery of Art.

Berrie, B. H., and S. Q. Lomax
1997. Azo pigments: Their history, synthesis, properties and use in artists' materials. In *Conservation Research 1996/1997*, 8–33. Monograph Series, no. 2, Studies in the History of Art, no. 57. Washington, D.C.: National Gallery of Art.

Besser, H., and J. Trant
1995. *Introduction to Imaging*. Santa Monica, Calif.: Getty Art History Information Program.

Bigger, S. W., and O. Delatycki
1989. The effect of hindered amine light stabilizers on the photooxidative stability of high-density polyethylene. *Journal of Polymer Science A: Polymer Chemistry* 27: 63–73.

Bigger, S. W., J. Scheirs, and O. Delatycki
1992. Effect of light intensity on the photooxidation kinetics of high-density polyethylene. *Journal of Polymer Science A: Polymer Chemistry* 30: 2277–80.

Bilz, M., and D. W. Grattan
1996a. The aging of parylene: Difficulties with the Arrhenius approach. In *11th Triennial Meeting, Edinburgh, Scotland, 1–6 September 1996: ICOM Committee for Conservation Preprints.* Vol. 2, 525–29. London: James & James.

1996b. The development of an apparatus for studying the effect of light exposure on museum materials. In *11th Triennial Meeting, Edinburgh, Scotland, 1–6 September 1996: ICOM Committee for Conservation Preprints.* Vol. 2, 559–65. London: James & James.

Bowman, J. G., and B. M. Reagan
1983. Filtered and unfiltered lights and their effects on selected dyed textiles. *Studies in Conservation* 28: 36–44.

Bradley, S., and V. Daniels
1990. Environment. In *A Guide to the Storage, Exhibition and Handling of Antiquities, Ethnographia and Pictorial Art.* Occasional Paper no. 66, ed. S. Bradley, 1–14. London: British Museum.

Brill, T. B.
1980. Decomposition of organic materials by light. Chap. 9 in *Light. Its Interaction with Art and Antiquities*, 173–95. New York: Plenum Press.

Bromelle, N. S., and P. Smith, eds.
1988. *Urushi: Proceedings of the Urushi Study Group, June 10–27, 1985, Tokyo.* Marina del Rey, Calif.: Getty Conservation Institute.

Bullock, L., and D. Saunders
1999. Measurement of cumulative exposure using Blue Wool standards. In *12th Triennial Meeting, Lyon, 29 August–3 September 1999: ICOM Committee for Conservation Preprints.* Vol. 1, 21–26. London: James & James.

Buschle-Diller, G., and S. H. Zeronian
1993a. Photodegradation of oxidized cotton cellulose. *Journal of Applied Polymer Science* 47: 1319–28.

1993b. Weathering and photodegradation of cellulose. Chap. 14 in *Photochemistry of Lignocellulosic Materials*, ed. C. Heitner and J. C. Scaiano, 177–89. ACS Symposium Series, no. 531. Washington, D.C.: American Chemical Society.

Cannon, J., M. Cannon, and G. Dalby-Quenet
1994. *Dye Plants and Dyeing*. Portland, Ore. Timber Press.

Cardamone, J. M., and W. N. Marmer
1994. Bleaching efficiency and photostability of wool, wool/cotton blends and all-cotton fabric. *Textile Chemist and Colorist* 26(6): 19–22.

Carlsson, D. J., L. H. Gan, and D. M. Wiles
1978. Photodegradation of aramids II: Irradiation in air. *Journal of Polymer Science: Polymer Chemistry Edition* 16: 2365–76.

Carpignano, R., P. Savarino, E. Barni, G. Di Modica, and S. S. Papa
1985. Developments in the application of quantitative structure–property relationships of dyes. *Journal of the Society of Dyers and Colourists* 101: 270–76.

Castellan, A., A. Nourmamode, S. Grelier, and P. Fornier de Violet
1996. Photocoloration du bois de pin maritime en présence d'additifs protecteurs. *Cellulose Chemistry and Technology* 30: 431–52.

Castellan, A., A. Nourmamode, C. Jaeger, and I. Forsskåhl
1993. Photochemistry of quinones and hydroquinones in solid 2-hydroxypropylcellulose films and on filter paper. In *Photochemistry of Lignocellulosic Materials*, ed. C. Heitner and J. C. Scaiano, 60–76. ACS Symposium Series, no. 53. Washington, D.C.: American Chemical Society.

Chatterjee, S., P. D. Davis, P. Gottschalk, M. E. Kurz, B. Sauerwein, X. Yang, and G. B. Schuster
1990. Photochemistry of carbocyanine alkyltriphenylborate salts: Intra-ion-pair electron transfer and the chemistry of boranyl radicals. *Journal of the American Chemical Society* 112: 6329–38.

Chemical and Engineering News
1994. Eastman Chemical markets new polyester homopolymer. *Chemical and Engineering News* 72(35): 9.

Chibisov, A. K., G. V. Zakharova, V. L. Shapovalov, A. I. Tolmachev, Yu. L. Briks, and Yu. L. Slominskii
1995. Flash photolysis of polymethine dyes with various counterions in nonpolar media. *High Energy Chemistry* 29: 192–98.

Choudhury, H., S. Collins, R. S. Davidson, and A. Castellan
1995. The photodegradation of lignin. In *Current Trends in Polymer Photochemistry*, ed. N. S. Allen, M. Edge, I. R. Bellobono, and E. Selli, 211–18. New York: Ellis Horwood.

Christie, R. M., W. T. Dryburgh, and P. N. Standring
1991. Some monoazoacetoacetanilide pigments derived from heterocyclic diazo components. *Dyes and Pigments* 16: 231–40.

Clayton, R. K.
1970. *Light and Living Matter*. Vol. 1, *The Physical Part*. New York: McGraw-Hill.

Colby, K. M.
1992. A suggested exhibition policy for works of art on paper. *Journal of the International Institute for Conservation–Canadian Group* 17: 3–12.

Cole, B. J. W., and K. V. Sarkanen
1987. Bleaching and brightness stabilization of high-yield pulps by sulfur-containing compounds. *Tappi Journal* 70(11): 117–22.

Collins, S., R. S. Davidson, and M. E. C. Hilchenbach
1994. The natural fluorescence of wool. *Dyes and Pigments* 24: 151–69.

Conservation Distribution List
1996. Search of various postings, bringing up "photoflash" and "UV safety" topics. ConsDistList 10:16, 10:17, 10:18, 10:20, 10:21. http://palimpsest.stanford.edu.

Crews, P. C.
1982. The influence of mordant on the lightfastness of yellow natural dyes. *Journal of the American Institute for Conservation* 21: 43–58.

1987. The fading rates of some natural dyes. *Studies in Conservation* 32: 65–72.

1988. A comparison of clear versus yellow ultraviolet filters in reducing fading of selected dyes. *Studies in Conservation* 33: 87–93.

1989. A comparison of selected UV filtering materials for the reduction of fading. *Journal of the American Institute for Conservation* 28: 117–25.

Crighton, J. S.
1992. Textile conservation. In *Polymers in Conservation*, ed. N. S. Allen, M. Edge, and M. V. Horie, 82–101. Cambridge: Royal Society of Chemistry.

Cuttle, C.
1988. Lighting works of art for exhibition and conservation. *Lighting Research and Technology* 20(2): 43–53.

Daniels, V.
1987. The blackening of vermilion by light. In *Recent Advances in the Conservation and Analysis of Artefacts*, comp. J. Black, 280–82. London: Institute of Archaeology Summer Schools Press.

1995. Factors influencing the wash-fastness of water colours. *The Paper Conservator* 19: 31–40.

Daniels, V., and B. Boyd
1986. The yellowing of thymol in the display of prints. *Studies in Conservation* 31: 156–58.

Davidson, R. S., G. M. Ismail, and D. M. Lewis
1987. The photosensitising properties and photostability of stilbene fluorescent whitening agents. *Journal of the Society of Dyers and Colourists* 103: 261–64.

Davidson, R. S., D. King, D. M. Lewis, and S. K. R. Jones
1985. The photostability of fluorescent whitening agents on wool. *Journal of the Society of Dyers and Colourists* 101: 291–94.

de la Rie, R.
1988a. Photochemical and thermal degradation of films of dammar resin. *Studies in Conservation* 33: 53–70.

1988b. Polymer stabilizers: A survey with reference to possible applications in the conservation field. *Studies in Conservation* 33: 9–22.

1993. Polymer additives for synthetic low-molecular-weight varnishes. In *10th Triennial Meeting, Washington, D.C, 22–27 August 1993: ICOM Committee for Conservation Preprints*. Vol. 2, 566–73. Paris: ICOM Committee for Conservation.

Derbyshire, A., and J. Ashley-Smith
1999. A proposed practical lighting policy for works of art on paper at the V&A. In *12th Triennial Meeting, Lyon, 29 August–3 September 1999: ICOM Committee for Conservation Preprints*. Vol. 1, 38–41. London: James & James.

Dessauer, R., and C. E. Looney
1989. Low amplification imaging systems. Chap. 8 in *Imaging Processes and Materials, Neblette's Eighth Edition,* ed. J. Sturge, V. Walworth, and A. Shepp, 276–77. New York: Van Nostrand Reinhold.

Dirckx, O., M. C. Triboulot-Trouy, N. Merlin, and X. Deglise
1992. Modifications de la couleur du bois d'*Abies grandis* exposé à la lumière solaire. *Annales des Sciences Forestières* (Annals of forest science) 49: 425–47.

Domergue, D. M., and J. F. Asmus
1976. Acceleration of embossed moleskin overpaint divestment through photodecomposition. In *Preprints of Papers Presented at the 4th Annual Meeting of the American Institute for Conservation of Historic and Artistic Works, Dearborn, Michigan, 29 May–01 June 1976,* 32–40. Washington, D.C.: American Institute for Conservation.

Douglas, P., and S. M. Townsend
1992. Photodegradation of photographic image dyes. In *The Imperfect Image—Photographs, Their Past, Present and Future: Conference Proceedings, Windermere, Cumbria, England, U.K.,* 274–86. London: Centre for Photographic Conservation.

Down, J. L.
1986. The yellowing of epoxy resin adhesives: Report on high-intensity light aging. *Studies in Conservation* 31: 159–70.

Dubois, A., M. Canva, A. Brun, F. Chaput, and J. P. Boilot
1996. Enhanced photostability of dye molecules trapped in solid xerogel matrices. *Synthetic Metals* 81: 305–8.

Duff, D. G., R. S. Sinclair, and D. Stirling
1977. Light-induced colour changes of natural dyes. *Studies in Conservation* 22: 161–69.

Duxbury, D. F.
1993. The photochemistry and photophysics of triphenylmethane dyes in solid and liquid media. *Chemical Reviews* 93: 381–433.

1994a. The sensitized fading of triphenylmethane dyes in polymer films. Part 1, *Dyes and Pigments* 25: 131–66.

1994b. The sensitized fading of triphenylmethane dyes in polymer films. Part 2, *Dyes and Pigments* 25: 179–204.

Eastman Kodak Co.
1990. Technical information about Kodak filters. *Kodak Photographic Filters Handbook,* 83–137. Rochester, NY: Eastman Kodak Co.

Egerton, G. S.
1948. The role of hydrogen peroxide in the photochemical degradation of cotton sensitized by vat dyes and some metallic oxides. *Textile Institute Journal. Part II: Transactions* 39: T305–18.

Egerton, G. S., and A. G. Morgan
1970a. The photochemistry of dyes I: Fundamental principles. *Journal of the Society of Dyers and Colourists* 86: 79–83.

1970b. The photochemistry of dyes II: Aspects of the fading process. *Journal of the Society of Dyers and Colourists* 86: 242–49.

Egerton, T. A., and C. J. King
1979. The influence of light intensity on photoactivity in TiO_2 pigmented systems. *Journal of the Oil and Colour Chemists' Association* 62: 386–91.

Ek, M., H. Lennholm, G. Lindblad, T. Iversen, and D. G. Gray
1993. Photochromic behavior of UV-irradiated mechanical pulps. In *Photochemistry of Lignocellulosic Materials*, ed. C. Heitner and J. C. Scaiano, 147–55. ACS Symposium Series, no. 531. Washington, D.C.: American Chemical Society.

Ellis, M. H.
1994. Drawings in fibre-tipped pen. New conservation challenges. In *Modern Works, Modern Problems? Conference Papers*, ed. A. Richmond, 114–21. Leigh, Worcester, England: Institute for Paper Conservation.

Ellis, M. H., and M. B. Yeh
1997. Categories of wax based drawing media. *WAAC (Western Association of Art Conservation) Newsletter* 19(3): 16–17.

Evans, N. A.
1980. Structural factors affecting light stability of dyed polymers. Chap. 3 in *Photochemistry of Dyed and Pigmented Polymers*, ed. N. S. Allen and J. F. McKellar, 93–159. London: Applied Science Publishers.

Evans, N. A., D. E. Rivett, and P. J. Waters
1976. Effect of photostability of 2-pyrazoline fluorescent whitening agents on the rate of yellowing of whitened wool. *Textile Research Journal* 46: 214–19.

Faucitano, A., A. Buttafava, G. Camino, and L. Greci
1996. Photo-oxidation and stabilization of polymers. *Trends in Polymer Science* 4(3): 92–98.

Feller, R. L.
1968. Studies on the darkening of vermilion by light. In *Report and Studies in the History of Art 1967*, 99–111. Washington, D.C.: National Gallery of Art.

1994. *Accelerated Aging: Photochemical and Thermal Aspects*. Marina del Rey, Calif.: Getty Conservation Institute.

Feller, R. L., M. Curran, and C. Bailie
1981. Photochemical studies of methacrylate coatings for the conservation of museum objects. In *Photodegradation and Photostabilization of Coatings*, ed. S. P. Pappas and F. H. Winslow, 183–96. Washington, D.C.: American Chemical Society.

1984. Identification of traditional organic colorants employed in Japanese prints and determination of their rates of fading. In *Japanese Woodblock Prints: A Catalogue of the Mary A. Ainsworth Collection*. Oberlin, Ohio: Allen Memorial Art Museum.

Feller, R. L., S. B. Lee, and J. Bogaard
1986. The kinetics of cellulose deterioration. In *Historic Textile and Paper Materials. Conservation and Characterization*, ed. H. L. Needles and S. H. Zeronian, 329–47. Advances in Chemistry Series, no. 212. Washington, D.C.: American Chemical Society.

Fireman, B.
1997. The technology and conservation of blueprints. *Paper Conservation News* (81): 3–5.

Fisher, G. J., C. Lewis, and D. Madill
1976. Laser flash photolysis of eosin and its complex with lysozyme. *Photochemistry and Photobiology* 24: 223–28.

Fornes, R. E., and R. D. Gilbert
1991. Environmental effects on polymers. In *Polymer and Fiber Science: Recent Advances*, 1–8. New York: VCH Publishers, Inc.

Fornes, R. E., R. D. Gilbert, B. S. Stowe, and G. P. Cheek
1973. Photodegradation of nylon 66 exposed to near UV radiation. *Textile Research Journal* 43: 713–15.

Forsskåhl, I., and C. Maunier
1993. Photocycling of chromophoric structures during irradiation of high-yield pulps. In *Photochemistry of Lignocellulosic Materials*, ed. C. Heitner and J. C. Scaiano, 156–66. ACS Symposium Series, no. 531. Washington, D.C.: American Chemical Society.

Forsskåhl, I., and H. Tylli
1993. Action spectra in the UV and visible region of light-induced changes of various refiner pulps. In *Photochemistry of Lignocellulosic Materials*, ed. C. Heitner and J. C. Scaiano, 45–59. ACS Symposium Series, no. 531. Washington, D.C.: American Chemical Society.

Fournier, J.-M., and P. L. Burnett
1994. Color rendition and archival properties of Lippmann photographs. *Journal of Imaging Science and Technology* 38: 507–12.

Fox, M. A., and M. T. Dulay
1996. Acceleration of secondary dark reactions of intermediates derived from adsorbed dyes on irradiated TiO_2 powders. *Journal of Photochemistry and Photobiology A: Chemistry* 98: 91–101.

Francis, R. C., C. W. Dence, T. C. Alexander, R. Agnemo, and S. Omori
1991. Photostabilization of thermomechanical pulps by alkylation and borohydride reduction. *Tappi Journal* 74(12): 127–33.

Frey, F., and R. Gschwind
1994. Mathematical bleaching models for photographic three-color materials. *Journal of Imaging Science and Technology* 38: 513–19.

Fujii, E., and H. Fujii
1992. Image stability evaluation of colour hard copies. In *The Imperfect Image—Photographs, Their Past, Present and Future: Conference Proceedings, Windermere, Cumbria, England, U.K.*, 294–300. London: Centre for Photographic Conservation.

Fujii, E., H. Fujii, and T. Hisanaga
1988. Evaluation on the stability of light faded images of color reversal films according to color difference in CIELab. *Journal of Imaging Technology* 14: 29–37.

Furuya, K., N. Furutachi, S. Oda, and K. Maruyama
1994. Photochemical reactions of 1*H*-pyrazolo[1,5-*b*][1,2,4]triazole azomethine dyes. *Journal of the Chemical Society, Perkin Transactions* 2: *Physical Organic Chemistry* 3: 531–36.

Gallotti, L.
1994. The action of light on dyes and on their textile substrates. Part 2, Photodegradation of dyes and natural fibers (in Italian). *Tinctoria* 91(4): 61–65.

General Electric Company
n.d. Photographic Lamp and Equipment Guide, P1.

Giacherio, D. J., and L. F. Valente
1985. Effects of wavelength, irradiation time, and coating composition on photochemical yellowing of model coatings. In *Second International Symposium: The Stability and Preservation of Photographic Images, August 25–28, 1985, the Public Archives of Canada and the Delta Ottawa Hotel, Ottawa, Ontario, Canada*, 51–52. Springfield, Va.: Society of Photographic Scientists and Engineers.

Giglio, K. D., D. B. Green, and B. Hutchinson
1995. Photocatalytic destruction of an organic dye using TiO_2 and solar energy. *Journal of Chemical Education* 72: 352–54, 625.

Giles, C. H., D. G. Duff, and R. S. Sinclair
1982. The relationship between dye structure and fastness properties. *Reviews of Progress in Coloration* 12: 58–65.

Giles, C. H., and S. D. Forrester
1980. Physical factors affecting the light stability of dyed and pigmented polymers. Chap. 2 in *Photochemistry of Dyed and Pigmented Polymers*, ed. N. S. Allen and J. F. McKellar, 51–91. London: Applied Science Publishers.

Giles, C. H., S. D. Forrester, R. Haslam, and R. Horn
1973. Lightfastness evaluation of colour photographs. *Journal of Photographic Science* 21: 19–23.

Giles, C. H., R. Haslam, and D. G. Duff
1976. An examination of the cause of the humidity sensitivity of the lightfastness of dyed materials. *Textile Research Journal* 46: 51–54.

Giles, C. H., B. J. Hojiwala, and C. D. Shah
1972. Quantum efficiency measurements of fading of some disperse dyes in nylon and polyester films and in solution. *Journal of the Society of Dyers and Colourists* 88: 403–7.

Giles, C. H., B. J. Hojiwala, C. D. Shah, and R. S. Sinclair
1974. Activation energies of light fading of non-ionic disperse dyes in hydrophobic polymers. *Journal of the Society of Dyers and Colourists* 90: 45–49.

Giles, C. H., D. P. Johari, and C. D. Shah
1968. Some observations on the kinetics of dye fading. *Textile Research Journal* 38: 1048–56.

Giles, C. H., and R. B. McKay
1963. The lightfastness of dyes: A review. *Textile Research Journal* 33: 528–77.

Giles, C. H., C. D. Shah, and D. Baillie
1969. Economical and efficient fading lamps. *Journal of the Society of Dyers and Colourists* 85: 410–17.

Giles, C. H., C. D. Shah, W. E. Watts, and R. S. Sinclair
1972. Oxidation and reduction in light fading of dyes. *Journal of the Society of Dyers and Colourists* 88: 433–35.

Goto, T., and T. Kondo
1991. Structure and molecular stacking of anthocyanins. Flower color variation. *Angewandte Chemie International English Edition* 30: 17–33.

Grassie, N., and G. Scott
1985. Photodegradation. Chap. 3 in *Polymer Degradation and Stabilisation.* Cambridge: Cambridge Univ. Press.

Green, L. R., and V. D. Daniels
1993. The use of N-methyl-2-pyrrolidone (NM2P) as a solvent for the analysis of indigoid dyes. *Dyes in History and Archaeology* 11: 10–18.

Gregory, P.
1994. Modern Reprographics. *Reviews of Progress in Coloration and Related Topics* 24: 1–16.

Grierson, S., D. G. Duff, and R. S. Sinclair
1985a. The colour and fastness of natural dyes of the Scottish highlands. *Journal of the Society of Dyers and Colourists* 101: 220–28.

1985b. Natural dyes of the Scottish highlands. *Textiles in History* 16(1): 23–43.

Griffiths, J.
1980. Solution and polymer photochemistry of azo dyes and related compounds. Chap. 6 in *Developments in Polymer Photochemistry—1*, ed. N. S. Allen, 145–90. London: Applied Science Publishers.

Hallas, G.
1979. The effects of terminal groups in 4-aminoazobenzene and disperse dyes related thereto. *Journal of the Society of Dyers and Colourists* 95: 285–94.

Hanlan, J. F.
1970. The effect of electronic photographic lamps on the materials of works of art. *Museum News* 48(10): 33–41.

Hasegawa, M., and K. Saigo
1990. Synthesis and properties of polymers having coumarin dimer moieties. In *Photochemistry and Photophysics.* Vol. 2, ed. J. F. Rabek and G. W. Scott, 27–56. Boca Raton, Fla.: CRC Press.

Hasegawa, Y.
1988. Urushi coating and color painting applied to Japanese architectural cultural monuments. In *Urushi: Proceedings of the Urushi Study Group, June 10–27, 1985, Tokyo*, ed. N. S. Bromelle and P. Smith, 57–67. Marina del Rey, Calif.: Getty Conservation Institute.

Havermans, J. B. G. A., and J. Dufour
1997. Photo oxidation of paper documents. *Restaurator* 18: 103–14.

Heitner, C.
1993a. Inhibition of light-induced yellowing of lignin-containing paper. In *Photochemistry of Lignocellulosic Materials,* ed. C. Heitner and J. C. Scaiano, 192–204. ACS Symposium Series, no. 531. Washington, D.C.: American Chemical Society.

1993b. Light-induced yellowing of wood-containing papers: An evolution of the mechanism. In *Photochemistry of Lignocellulosic Materials,* ed. C. Heitner and J. C. Scaiano, 2–25. ACS Symposium Series, no. 531. Washington, D.C.: American Chemical Society.

Heller, A., Y. Degani, D. W. Johnson Jr., and P. K. Gallagher
1987. Controlled suppression and enhancement of the photoactivity of titanium dioxide (rutile) pigment. *Journal of Physical Chemistry* 91: 5987–91.

Hemmendinger, H.
1972. Mechanisms of release of energy absorbed by dyes. *Journal of Color and Appearance* 1(6): 11–14, 29.

Herbst, W., and K. Hunger
1993. *Industrial Organic Pigments.* Weinheim, Germany: VCH Publishers, Inc.

Hodak, J., C. Quinteros, M. I. Litter, and E. San Román
1996. Sensitization of TiO_2 with phthalocyanins. Part 1, Photooxidations using hydroxoaluminum tricarboxymonoamine-phthalocyanine adsorbed on TiO_2. *Journal of the Chemical Society, Faraday Transactions* 92: 5081–88.

Hoffmann, E., and A. Saracz
1969. Weathering of paint films I: Chalking caused by zinc oxide in latex paints. *Journal of the Oil and Colour Chemists Association* 52: 113–32.

Hon, D. N.-S.
1991. Photochemistry of wood. Chap. 11 in *Wood and Cellulosic Chemistry*, ed. D. N.-S. Hon and N. Shiraishi, 525–55. New York: Marcel Dekker.

Horie, C. V.
1989. Polymers derived from cellulose. Chap. 10 in *Materials for Conservation*, 124–34. Oxford: Butterworth/Heinemann.

1990. Fading of feathers by light. In *9th Triennial Meeting, Dresden, German Democratic Republic, 26–31 August 1990: ICOM Committee for Conservation Preprints.* Vol. 2, 431–36. Los Angeles: ICOM Committee for Conservation.

Hoyle, C. E., D. Creed, P. Subramanian, I. B. Rufus, P. Chatterton, M. Bahadur, and P. Wisian-Neilson
1994. Photophysics and photochemistry of poly(methylphenylphosphazene) and poly(methylphenylphosphazene)-graft-polystyrene copolymers. In *Inorganic and Organometallic Polymers II: Advanced Materials and Intermediates*, ed. P. Wisian-Neilson, H. R. Allcock, and K. J. Wynne, 324–42. ACS Symposium Series, no. 572. Washington, D.C.: American Chemical Society.

Hoyle, C. E., H. Shah, and K. Moussa
1996. Photolysis of methylene 4,4′-diphenyldiisocyanate-based polyurethane ureas and polyureas. In *Polymer Durability, Degradation, Stabilization and Lifetime Prediction*, ed. R. L. Clough, N. C. Billingham, and K. T. Gillen, 91–111. Advances in Chemistry Series, no. 249. Washington, D.C.: American Chemical Society.

Imagi, K., N. Ikeda, H. Masuhara, M. Nishigaki, and M. Isogawa
1987. Photochemical transient species of poly(ethylene terephthalate) powders as

revealed by the diffuse reflectance laser photolysis method. *Polymer Journal* 19: 999–1001.

Ioanovici, E.
1975. Degradarea operelor de pictura in timpul filmarilor si fotografierilor (Degradation of paintings during the making of motion pictures or photographs; English abstract in *Art and Archaeology Technical Abstracts,* no. 13-367, 1976). *Revista Muzeelor Si Monumentelor* (2): 39–45.

Jagger, J.
1967. Appendix A: Tables. In *Introduction to Research in Ultraviolet Photobiology,* 129–32. Englewood Cliffs, N.J.: Prentice Hall.

1977. Phototechnology and biological experimentation. In *The Science of Photobiology*, ed. K. C. Smith, 1–26. New York: Plenum Publishing Co.

Jarry, N.
1996. Computer imaging technology: The process of identification. *Book and Paper Group Annual* 15: 53–60.

Jeon, K. S., and I. Block
1990. Photodegradation of cellulosics. Part 1, Effects of temperature and humidity on tear strength retention. In *9th Triennial Meeting, Dresden, German Democratic Republic, 26–31 August 1990: ICOM Committee for Conservation Preprints.* Vol. 1, 302–6. Los Angeles: ICOM Committee for Conservation.

Jockusch, S., H.-J. Timpe, W. Schnabel, and N. J. Turro
1996. Photoreduction of organic dyes in ketone amine systems. *Journal of Photochemistry and Photobiology A: Chemistry* 96: 129–36.

Johnson, D. G., and B. M. Reagan
1990. Influence of antimicrobial agents on dye fading and fiber yellowing in nylon. *Textile Chemist and Colorist* 22(4): 21–24

Johnson, L. D., W. D. Tincher, and H. C. Bach
1969. Photodegradative wavelength dependence of thermally resistant organic polymers. *Journal of Applied Polymer Science* 13: 1825–32.

Johnston-Feller, R., R. L. Feller, C. W. Bailie, and M. Curran
1984. The kinetics of fading: Opaque paint films pigmented with alizarin lake and titanium dioxide. *Journal of the American Institute for Conservation* 23: 114–29.

Jones, G. II, Z. Feng, and C. Oh
1995. Photoinduced electron transfer for an eosin-tyrosine conjugate: Activity of the tyrosinate anion in long-range electron transfer in a protein-like polymer matrix. *Journal of Physical Chemistry* 99: 3883–88.

Jones, G. II, C. Oh, and K. Goswami
1991. The photochemistry of triarylmethane dyes bound to polyelectrolytes: Photoinduced electron transfer involving bound dye monomers and dimers. *Journal of Photochemistry and Photobiology A: Chemistry* 57: 65–80.

Jones, N. M. M.
1990. Archival copies of Thermofax, Verifax, and other unstable records. Appendix in *National Archives Technical Information Paper No. 5.* Washington, D.C.: National Archives and Records Administration.

Jouan, X., and J.-L. Gardette
1991. Photooxidation of ABS at long wavelengths. *Journal of Polymer Science, Polymer Chemistry Edition* 29: 685–96.

Kajitani, N.
1989. Conservation maintenance of tapestries at the Metropolitan Museum of Art, 1987. In *The Conservation of Tapestries and Embroideries: Proceedings of Meetings at the Institute Royal du Patrimoine Artistique Brussels, Belgium*, 53–65. Marina del Rey, Calif.: Getty Conservation Institute.

Kamat, P. V.
1993. Photochemistry on nonreactive and reactive (semiconductor) surfaces. *Chemical Reviews* 93: 267–300.

Kamel, M., S. A. Amin, M. S. Afifi, and M. Tera
1979. Relationship between lightfastness and the physical properties of simple monoazo dyestuffs. *Textile Research Journal* 49: 260–61.

Kaneko, M., and A. Yamada
1981. Solid phase photoreduction of methylviologen on cellulose. *Macromolecular Chemistry* 182: 1111–18.

Katsuda, N., S. Yabushita, K. Otake, T. Omura, and T. Takagishi
1996. Photodegradation of a disperse dye on polyester fiber and in solution. *Dyes and Pigments* 31: 291–300.

Keller, K.
1993. Photographic properties and their measurement. Sec. 6.4 in *Science and Technology of Photography*. Weinheim, Germany: VCH Publishers, Inc.

Kelly, G. P., M. Mollah, and F. Wilkinson
1990. A study of transient absorption in wool keratin and silk fibroin by diffuse reflectance laser-flash photolysis. *Journal of the Textile Institute* 81: 91–94.

Kenjo, T.
1987. Discoloration of red colors irradiated with some monochromatic lights (in Japanese; English abstract in *Art and Archaeology Technical Abstracts*, no. 25-492, 1988). *Hozon Kagaku* (Science for Conservation) 26: 31–34.

1988. Discoloration and deterioration of certain wooden materials irradiated by the sunlight (in Japanese; English abstract in *Art and Archaeology Technical Abstracts*, no. 25-1800). In *Conservation-Restoration of Leather and Wood, Training of Restorers: Sixth International Restorer Seminar, 1987*, 29–30. Budapest: National Centre of Museums.

Kenney, A. R., and S. Chapman
1996. Technical overview. Chap. 2 in *Digital Imaging for Libraries and Archives*. Ithaca, N.Y.: Cornell Univ. Press.

Kenney, A. R., and L. K. Personius
1992. Digital capture, paper facsimiles, and network access. *Cornell/Xerox/Commission on Preservation and Access/Joint Study in Digital Preservation, Report: Phase1, January 1990–December 1991*. Washington, D.C.: Commission on Preservation and Access.

Kenzo [Kenjo], T.
1986. Studies on films of Japanese lacquer. Part 3, The change in infrared spectra of Japanese lacquer films in the course of their hardening and deterioration, and discus-

sion on conservation of Japanese lacquered art objects. Trans. Bureau of Translation of the Secretary of State's Office, Canada. Ottawa, Ontario.

Kim, B. Y., A. Isogai, F. Onabe, and M. Usuda
1989. Photochemical reactions of paper. Chap. 50 in *Wood Processing and Utilization*, ed. J. F. Kennedy, G. O. Phillips, and P. A. Williams, 382–86. Chichester, England: Ellis Horwood, Ltd.

Kirby, J.
1993. Fading and colour change of Prussian blue: Occurrences and early reports. *National Gallery Technical Bulletin* 14: 63–71.

Kisch, H., and R. Künneth
1991. Photocatalysis by semiconductor powders. Preparative and mechanistic aspects. Chap. 2 in *Photochemistry and Photophysics*. Vol. 4, ed. J. F. Rabek, 131–75. Boca Raton, Fla.: CRC Press.

Klemchuk, P. P.
1983. Influence of pigments on the light stability of polymers: A critical review. *Polymer Photochemistry* 3: 1–17.

Kosar, J.
1965. Inorganic compounds. Chap. 1 in *Light Sensitive Systems*. New York: John Wiley.

Kovářová, J., J. Rotschová, O. Brede, and M. Burgers
1995. The effect of transformation products of the antioxidant BHT on the initial stages of thermo- and photo-oxidation of LDPE. *Canadian Journal of Chemistry* 73: 1862–68.

Kramer, H. E. A.
1986. The lightfastness or the non-photochemistry of dyes. *Chimia* 40:160–69.

Kühn, H.
1986. Zinc white. In *Artists' Pigments: A Handbook of Their History and Characterization*. Vol. 1, ed. R. L. Feller, 169–86. Washington, D.C.: National Gallery of Art.

Kühn, H., and M. Curran
1986. Chrome yellow and other chromate pigments. In *Artists' Pigments: A Handbook of Their History and Characterization*. Vol. 1, ed. R. L. Feller, 187–217. Washington, D.C.: National Gallery of Art.

Kumanotani, J.
1988. The chemistry of Oriental lacquer (*Rhus verniciflua*). In *Urushi: Proceedings of the Urushi Study Group, June 10–27, 1985, Tokyo*, ed. N. S. Bromelle and P. Smith, 243–51. Marina del Rey, Calif.: Getty Conservation Institute.

Kuramoto, N., and T. Kitao
1979. Contribution of singlet oxygen to the photofading of indigo. *Journal of the Society of Dyers and Colourists* 95: 257–61.

1981. Mechanism of the photofading of dye: Contribution of singlet oxygen to the photofading of aminoanthraquinone dyes. *Dyes and Pigments* 2: 133–42.

1982. The contribution of singlet oxygen to the photofading of triphenylmethane and related dyes. *Dyes and Pigments* 3: 49–58.

Kuruppillai, R. V., S. P. Hersh, and P. A. Tucker
1986. Degradation of silk by heat and light. In *Historic Textile and Paper Materials. Conservation and Characterization*, ed. H. L. Needles and S. H. Zeronian, 111–27. Advances in Chemistry Series, no. 212. Washington, D.C.: American Chemical Society.

Lacoste, J., Y. Israëli, and J. Lemaire
1996. Photoaging of substituted and unsubstituted silicones. In *Polymer Durability. Degradation, Stabilization and Lifetime Prediction*, ed. R. L. Clough, N. C. Billingham, and K. T. Gillen, 77–89. Advances in Chemistry Series, no. 249. Washington, D.C.: American Chemical Society.

Ladisch, C. M., R. R. Brown, and K. B. Showell
1983. Photodegradation of reactive dyed cotton. *Textile Chemist and Colorist* 15: 209–12.

Lafontaine, R. H.
1979. The lightfastness of felt-tip pens. *Journal of the International Institute for Conservation–Canadian Group* 4(1): 9–16.

Larsen, R., M. Vest, and U. B. Kejser, eds.
1994. STEP Leather Project. Evaluation of the correlation between natural and artificial ageing of vegetable tanned leather and determination of parameters for standardization of an artificial ageing method. *Research Report No. 1*. Copenhagen: Bjarnholt Repro.

Lavédrine, B., M. Gillet, and C. Garnier
1999. Mise au point d'un actinomètre pour le contrôle de l'exposition des photographies et des objets sensibles à la lumière. In *12th Triennial Meeting, Lyon, 29 August–3 September 1999: ICOM Committee for Conservation Preprints*. Vol. 1, 65–69. London: James & James.

Leaver, I. H.
1978. Sensitized photooxidation of wool by fluorescent whitening agents. A direct spectral study. *Photochemistry and Photobiology* 27: 451–56.

Leaver, I. H., and B. Milligan
1984. Fluorescent whitening agents—a survey, 1974–1982. *Dyes and Pigments* 5: 109–44.

Leclerc, F., and F. Flieder
1992. Influence of optical brighteners on paper permanence. In *Papers Presented to the Third International Institute of Paper Conservation Conference at the University of Manchester, Institute of Science and Technology, 1–4 April 1992*, ed. S. Fairbrass, 257–63. Leigh, Worcester, England: Institute for Paper Conservation.

Lee, S. B., J. Bogaard, and R. L. Feller
1989. Darkening of paper following exposure to visible and near-ultraviolet radiation. *Journal of the American Institute for Conservation* 28: 1–18.

1994. Bleaching by light I: Effect of pH on the bleaching or darkening of papers in the dry and in the immersed condition under visible and near-ultraviolet radiation. In *Conservation of Historic and Artistic Works on Paper, Proceedings of Symposium 88*, ed. H. D. Burgess, 181–90. Ottawa: Canadian Conservation Institute.

Lemaire, J., J.-L. Gardette, B. Mailhot, and X. Jouan
1995. Weathering relevant photooxidation mechanisms of styrenic polymers. Chap. 12 in *Current Trends in Polymer Photochemistry*, ed. N. S. Allen, M. Edge, I. R. Bellobono, and E. Selli, 175–99. New York: Ellis Horwood.

Levison, H. W., F. Sutil, and E. T. Vanderbrink
1987. Lightfastness of pigmented handmade papers. *Color Research and Application* 12: 37–41.

Lewis, D. M.
1989. Some aspects of the photochemistry of fibrous proteins. *Colourage* 36: 25–31.

Liles, J. N.
1990. *The Art and Craft of Natural Dyeing: Traditional Recipes for Modern Use.* Knoxville, Tenn.: Univ. of Tennessee Press.

Lishan, D. G., P. W. Harris, K. Brahim, and R. L. Jackson
1988. Photodegradation of acid dyes in nylon films. *Journal of the Society of Dyers and Colourists* 104: 33–37.

Liu, D., G. L. Hug, and P. V. Kamat
1995. Photochemistry on surfaces: Intermolecular energy and electron transfer processes between $Ru(bpy)_3^{2+}$ and H-aggregates of cresyl violet on SiO_2 and SnO_2 colloids. *Journal of Physical Chemistry* 99: 16768–75.

Lowe, A., C. Franklin, W. Owen, and B. Smith
1997. *Digital Prints.* Bristol, England: Permaprint, Senecio Press, and the Print Research Unit of the University of the West of England.

Mailly, V., J. F. Le Nest, J. M. Serra Tosio, and J. Silvy
1997. Yellowing of coated papers under the action of heat, daylight radiation and nitrogen gas. *Tappi Journal* 80(5): 176–83.

Mansour, M. P., H. J. Cornell, and I. A. Holt
1988. Photoyellowing of wool pretreated with chlorinated solvents. *Textile Research Journal* 58: 246.

Marcandalli, B., A. Seves, E. Dubini-Paglia, and P. L. Beltrame
1990. Photochemical *trans-cis* isomerization of some 4-diethylaminoazobenzenes. *Dyes and Pigments* 14: 79–88.

March, J.
1985. Photochemistry. Chap. 7 in *Advanced Organic Chemistry: Reactions, Mechanisms and Structure*, 3rd ed., 202–17. New York: John Wiley & Sons.

Martin, K. G., and R. I. Tilley
1971. Influence of radiation wavelength on photo-oxidation of unstabilised PVC. *British Polymer Journal* 3: 36–40.

Massafra, M. R., E. Selli, S. Salsa, and B. Marcandalli
1999. Kinetic study on the sunlight-induced degradation of acid azo dyes on silk. *Dyes and Pigments* 40: 171–80.

Matsushima, R., H. Mizuno, and H. Itoh
1995. Photochromic properties of 4′-amino-substituted 2-hydroxychalcones. *Journal of Photochemistry and Photobiology A: Chemistry* 89: 251–56.

Mazet, J. F., M. C. Triboulot-Trouy, A. Merlin, G. Janin, and X. Deglise
1993. Modification de couleur du bois de chênes européens exposés à la lumière solaire. *Annales des Sciences Forestières* (Annals of forest science) 50: 119–46.

McAlpine, E., and R. S. Sinclair
1977. The light stability of 1,4-disubstituted aminoanthraquinone dyes; the influence of substituent and substrate. *Textile Research Journal* 47: 283–89.

McCrady, E.
1998. The nature of permanence. Selections from "North American Permanent Papers." *The Abbey Newsletter*. Available online from Abbey Publications at http://palimpsest.stanford.edu/byorg/abbey/napp/perman.html.

McElhone, J. P.
1992. Determining responsible display conditions for photographs. In *The Imperfect Image—Photographs, Their Past, Present and Future: Conference Proceedings, Windermere, Cumbria, England, U.K.*, 182–92. London: Centre for Photographic Conservation.

McKellar, J. F., and N. S. Allen
1979. *Photochemistry of Man-Made Polymers*. London: Applied Science Publishers.

McLaren, K.
1956. The spectral regions of daylight which cause fading. *Journal of the Society of Dyers and Colourists* 72: 86–99.

1983. Photophysics and photochemistry of colouring matters. In *The Colour Science of Dyes and Pigments*, 37–53. Bristol, England: Adam Hilger.

McNeill, I. C.
1992. Fundamental aspects of polymer degradation. In *Polymers in Conservation*, ed. N. S. Allen, M. Edge, and M. V. Horie, 14–31. Cambridge: Royal Society of Chemistry.

Mehta, H. P., and A. T. Peters
1981. Substituent effects on the colour, dyeing and fastness properties of 4-*N*-β-cyanoethyl-*N*-β-hydroxyethylaminoazobenzenes. Part 1—Monosubstituted derivatives. *Dyes and Pigments* 2: 259–69.

1982. Substituent effects on the colour, dyeing and fastness properties of 4-*N*-β-cyanoethyl-*N*-β-hydroxyethylaminoazobenzenes. Part 2—Disubstituted derivatives. *Dyes and Pigments* 3: 71–78.

Messier, P.
1997. Laser digitization and rapid prototyping. *Conservation Distribution List* 11: 50. Available online at http://palimpsest.stanford.edu.

Michalski, S.
1987. Damage to museum objects by visible radiation (light) and ultraviolet radiation (UV). In *Lighting: A Conference on Lighting in Museums, Galleries and Historic Houses*, ed. Museums Association, 3–15. Bristol, England: Museums Association, UKIC, and Group of Designers and Interpreters in Museums.

1994. A systematic approach to preservation: Description and integration with other museum activities. In *Preventive Conservation: Practice, Theory and Research: Preprints of the Contributions to the Ottawa Congress*, ed. A. Roy and P. Smith, 8–11. London: International Institute for Conservation.

Milligan, B., and L. A. Holt
1974. Fluorescent whitening agents I, *bis*-4,4′-(4″-methoxy-6″-phenoxy-*s*-triazin-2″ylamino)stilbene-2,2′-disulfonic acid: Its photodecomposition in solution and on wool. *Australian Journal of Chemistry* 27: 195–203.

Mills, J. S., and R. White

1994a. Dyestuffs and other coloured materials. Chap. 10 in *The Organic Chemistry of Museum Objects*, 2d ed., 141–69. London: Butterworth/Heinemann.

1994b. Natural resins and lacquers. Chap. 8 in *The Organic Chemistry of Museum Objects*, 2nd ed., 83–110. London: Butterworth/Heinemann.

Mizrachi, S.

1994. Some aspects of the conservation of pigment-coated papers. In *Modern Works, Modern Problems? Conference Papers*, ed. A. Richmond, 41–49. Leigh, Worcester, England: Institute for Paper Conservation.

Móger, G., G. Köhler, and N. Getoff

1996. Enhanced photochemical degradation of the haematoporphyrin dication irradiated at 560 nm in a lipid-like environment. *Journal of Photochemistry and Photobiology B: Biology* 33: 27–37.

Moor, I. L., and A. H. Moor

1992. Exhibiting photographs: The effect of the exhibition environment on photographs. In *The Imperfect Image—Photographs, Their Past, Present and Future: Conference Proceedings, Windermere, Cumbria, England, U.K.*, 193–201. London: Centre for Photographic Conservation.

Mortensen, A., and L. H. Skibsted

1996. Kinetics of photobleaching of β-carotene in chloroform and formation of transient carotenoid species absorbing in the near infrared. *Free Radical Research* 25: 355–68.

Moura, J. C. V. P., A. M. F. Oliveira-Campos, and J. Griffiths

1997. The effect of additives on the photostability of dyed polymers. *Dyes and Pigments* 33: 173–96.

Naguib, Y. M. A., C. Steel, S. G. Cohen, and M. A. Young

1996. Triplet-sensitized photobleaching of crystal violet. *Journal of Photochemistry and Photobiology A: Chemistry* 96: 149–54.

Nansheng, D., W. Feng, T. Shizhong, and F. Tao

1997. Photodegradation of dyes in aqueous solutions containing Fe(III)-hydroxy complex. II: Solar photodegradation kinetics. *Chemosphere* 34: 2725–35.

Nasr, C., D. Liu, S. Hotchandani, and P. V. Kamat

1996. Dye-capped semiconductor nanoclusters: Excited state and photosensitization aspects of rhodamine 6G H-aggregates bound to SiO_2 and SnO_2 colloids. *Journal of Physical Chemistry* 100: 11054–61.

Nasu, Y., F. Nakazawa, and M. Kashiwagi

1985. Emission and excitation spectra of natural yellow dyes on silk fabrics before and after fading. *Journal of the Society of Dyers and Colourists* 101: 173–76.

Needles, H. L., V. Cassman, and M. J. Collins

1986. Mordanted, natural-dyed wool and silk fabrics. Light and burial-induced changes in the color and tensile properties. In *Historic Textile and Paper Materials: Conservation and Characterization*, ed. H. L. Needles and S. H. Zeronian. Advances in Chemistry Series, no. 212, 199–210. Washington, D.C.: American Chemical Society.

Neevel, J. G.
1994. Exposure of objects of art and science to light from electronic flash-guns and photocopiers. In *Contributions of the Central Research Laboratory to the Field of Conservation and Restoration*, ed. H. Verschoor and J. Mosk, 77–87. Amsterdam: Central Research Laboratory.

Neevel, J. G., H. C. A. van Beek, and B. van de Graaf
1992. Flash photolysis of azo dyes in aqueous solutions of biacetyl. *Journal of the Society of Dyers and Colourists* 108: 150–54.

Nelson, G. M., and Z. W. Wicks
1982. Polymers of N-butyl methacrylate with covalently bound U.V. stabilizer. In *Resins in Conservation: Proceedings of the Symposium, Edinburgh, 1982*, ed. N. H. Tennent and J. H. Townsend, 4.1–4.13. Edinburgh: Scottish Society for Conservation and Restoration.

Nicholls, C. H.
1980. Photodegradation and photoyellowing of wool. Chap. 5 in *Developments in Polymer Photochemistry—I*, ed. N. S. Allen, 125–43. London: Applied Science Publishers.

Niu, E., K. P. Ghiggino, A. W.-H. Mau, and W. H. F. Sasse
1988. Fluorescence and photochemistry of dye sensitizers in Nafion membrane. *Journal of Luminescence* 40 & 41: 563–64.

Norville-Day, H.
1994. The conservation of faxes and colour photocopies, with special reference to David Hockney's "home-made" prints. In *Modern Works, Modern Problems? Conference Papers*, ed. A. Richmond, 66–72. Leigh, Worcester, England: Institute for Paper Conservation.

Oda, H., and T. Kitao
1985. Intramolecular quenching of the photofading of some dyes. *Journal of the Society of Dyers and Colourists* 101: 177–79.

1990. Role of counter-ions in the photofading reaction of crystal violet lactone. *Dyes and Pigments* 12: 97–105.

Oda, H., N. Kuramoto, and T. Kitao
1981. Mechanism of the photofading of stilbene-fluorescent dyes. Contribution of singlet oxygen to the photofading of 4,4′-bisacetamidostilbene. *Journal of the Society of Dyers and Colourists* 97: 462–64.

Oddy, A.
1996. The Forbes Prize lecture 1996. *International Institute for Conservation (IIC) Bulletin* 1996(5): 1–6.

Ogbobe, O., and N. N. Ossai
1992. Effect of colour on the mechanical and chemical properties of naturally photooxidized high density polyethylene. *Acta Polymerica* 43: 173–76.

Oger, B.
1996. Fastness to light and washing of direct dyes for cellulosic textiles. *Studies in Conservation* 41: 129–35.

Ohtani, B., S. Adzuma, S.-I. Nishimoto, and T. Kagiya
1992. Photocatalytic degradation of polyethylene film by incorporated extra-fine particles of titanium dioxide. *Polymer Degradation and Stability* 35: 53–60.

Okada, Y., M. Hirose, T. Kato, H. Motomura, and Z. Morita
1990a. Fading of vinylsulfonyl reactive dyes on cellulose in admixture under wet conditions. *Dyes and Pigments* 14: 265–85.

1990b. Photofading of vinylsulfonyl reactive dyes on cellulose under wet conditions. *Dyes and Pigments* 14: 113–27.

Okada, Y., T. Kato, H. Motomura, and Z. Morita
1990. Catalytic fading of vinylsulfonyl reactive dye mixtures on cellulose under wet conditions. *Dyes and Pigments* 12: 197–211.

Orlenko, K., and E. Stewart
1998. Conservation implications of computer generated printing. In *IPC Conference Papers London 1997: Proceedings of the Fourth International Conference of the Institute of Paper Conservation, 6–9 April 1997*, ed. J. Eagan, 166–74. Leigh, Worcester, England: Institute of Paper Conservation.

Oye, R.
1989. Degradation of bookpaper. Chap. 49 in *Wood Processing and Utilization*, ed. J. F. Kennedy, G. O. Phillips, and P. A. Williams, 373–79. Chichester, England: Ellis Horwood, Ltd.

Padfield, T., and S. Landi
1966. The light-fastness of the natural dyes. *Studies in Conservation* 11: 181–96.

Pegram, J. E., and A. L. Andrady
1991. Outdoor weathering of polystyrene foam. Chap. 22 in *Polymer and Fiber Science: Recent Advances*, ed. R. E. Fornes and R. D. Gilbert. New York: VCH Publishers, Inc.

Perkinson, R.
1974. On conservation: The problem of lighting works of art on paper. *Museum News* 53(3): 5–7.

Phillips, G. O., and J. C. Arthur Jr.
1985. Photochemistry and radiation chemistry of cellulose. In *Cellulose Chemistry and Its Applications*, ed. T. P. Nevell and S. H. Zeronian, 290–311. Chichester, England: Ellis Horwood, Ltd.

Phillips, R.
1983a. Incandescent sources. Chap. 5 in *Sources and Applications of Ultraviolet Radiation*, 162–69. London: Academic Press.

1983b. Xenon lamps. Chap. 11 in *Sources and Applications of Ultraviolet Radiation*, 299–332. London: Academic Press.

Photocopying and laser printing processes—their stability and permanence
1993. *Astralian Archives* (September): 1–3.

Pickett, J. E.
1990. Photodegradation and stabilisation of PPO® resin blends. In *Mechanisms of Polymer Degradation and Stabilisation*, ed. G. Scott, 135–67. London: Elsevier.

Porter, G.
1986. Forty years of photochemistry. *Journal of the Chemical Society, Faraday Transactions 2: Molecular and Chemical Physics* 82: 2445–51.

Preservation photocopying in libraries and archives
1987. Proceedings of the Conference of the U. S. National Archives and Records Administration, Washington, D.C. *Restaurator* 8(1).

Pretzel, B.
1992. Analysis of comparative colour changes occurring in a set of 19th century photographs by Lady Hawarden. In *The Imperfect Image—Photographs, Their Past, Present and Future: Conference Proceedings, Windermere, Cumbria, England, U.K.*, 165–81. London: Centre for Photographic Conservation.

Quye, A. J. Wouters, and J. J. Boon
1996. A preliminary study of light-ageing effects on the analysis of natural flavonoid-dyed wools by photodiode array HPLC and by direct temperature mass spectrometry. In *11th Triennial Meeting, Edinburgh, Scotland, 1–6 September 1996: ICOM Committee for Conservation Preprints.* Vol. 2, 704–13. London: James & James.

Rabek, J. F.
1987a. Electronically excited states. Chap. 1 in *Mechanisms of Photophysical Processes and Photochemical Reactions in Polymers.* Chichester, England: John Wiley & Sons.

1987b. Photodegradation of polymers. Chap. 14, sec. 14.3 in *Mechanisms of Photophysical Processes and Photochemical Reactions in Polymers,* 510–20. Chichester, England: John Wiley & Sons.

Rånby, B.
1989. Photodegradation and photo-oxidation of cellulose and cellulose derivatives. In *Wood Processing and Utilization*, ed. J. F. Kennedy, G. O. Phillips, and P. A. Williams, 353–60. Chichester, England: Ellis Horwood, Ltd.

1993. Basic reactions in the photodegradation of some important polymers. *Pure and Applied Chemistry* A30(9/10): 583–94.

1995. Photoinitiated modifications of synthetic polymers: Photocrosslinking and surface photografting. Chap. 2 in *Current Trends in Polymer Photochemistry*, ed. N. S. Allen, M. Edge, I. R. Bellobono, and E. Selli, 23–39. New York: Ellis Horwood.

Rånby, B., and J. F. Rabek
1992. Photodegradation of polymer materials. Chap. 12 in *Comprehensive Polymer Science,* ed. G. Allen. First Supplement, ed. S. L. Aggarwal and S. Russo, 253–83. Oxford: Pergamon Press.

Ray, A., and G. M. Deheri
1995. CNDO/II study on the fading of anthraquinone dyes. *Dyes and Pigments* 27: 327–31.

Reagan, B. M., A. Sarmadi, E. P. Easter, P. C. Crews, C. M. Ladisch, D. G. Johnson, A. Dale, J. M. Laughlin, C. L. Orr, and R. K. Lattie
1989. Influence of fluorescent whitening agents on silk phototendering and dye fading. In *Book of Papers: 1989 International Conference and Exhibition*, 37–42. Research Triangle Park, N.C.: American Association of Textile Chemists and Colorists.

Reine, A. H., and J. C. Arthur Jr.
1970. Photochemistry of cotton cellulose: The direct action of near-ultraviolet light on purified fibrous cotton cellulose. *Textile Research Journal* 40: 90–92.

Reinert, G.
1988. Photostability of polyamide fibers. *Melliand Textilberichte* 69: E29–32 (English translation of German ed.).

Rembold, M. W., and H. E. A. Kramer
1978. Singlet oxygen as an intermediate in the catalytic fading of dye mixtures. *Journal of the Society of Dyers and Colourists* 94: 12–17.

1980. The role of anthraquinoid dyes in the "catalytic fading" of dye mixtures—substituent-dependent triplet yield of diaminoanthraquinones. *Journal of the Society of Dyers and Colourists* 96: 122–26.

Robert, S., and C. Daneault
1995. Yellowing mechanism and kinetics of thick handsheets of softwood thermomechanical pulp. *Journal of Wood Chemistry and Technology* 15(1): 113–33.

Roberts, D. A.
1997. A guide to speaking the language of radiometry and photometry. In *The Photonics Directory.* Book 3, *The Photonics Design and Applications Handbook, 43rd International Edition*, H72–75. Pittsfield, Mass.: Laurin Publishing Co.

Roberts, J. D., R. Stewart, and M. C. Caserio
1971. Photochemistry. Chap. 26 in *Organic Chemistry: Methane to Macromolecules*, 695–709. New York: W. A. Benjamin, Inc.

Rusznák, I., J. Frankl, and J. Gombkötő
1985. Photoreactivity of wool dyed with reactive dyes. *Journal of the Society of Dyers and Colourists* 101: 130–36.

Saito, M., C. Minemura, N. Nanashima, and M. Kashiwagi
1988. Color fading behavior of anthraquinone dyes due to environmental conditions. *Textile Research Journal* 58: 450–54.

Sancho-Arroyo, M., and J.-P. Rioux
1996. L'utilisation au musée des équipements à éclairs électroniques. *Techne* 4: 128–36.

San Román, E.
1996. Immobilized phthalocyanines as red-light photosensitizers. *Journal of Photochemistry and Photobiology A: Chemistry* 102: 109–12.

Saunders, D.
1995. Photographic flash: Threat or nuisance? *National Gallery Technical Bulletin* 16: 66–72.

Saunders, D., and J. Kirby
1994a. Light-induced colour changes in red and yellow lake pigments. *National Gallery Technical Bulletin* 15: 79–97.

1994b. Wavelength-dependent fading of artists' pigments. In *Preventive Conservation: Practice, Theory and Research*, ed. A. Roy and P. Smith, 190–94. London: International Institute for Conservation.

1996. Light-induced damage: Investigating the reciprocity principle. In *11th Triennial Meeting, Edinburgh, Scotland, 1–6 September 1996: ICOM Committee for Conservation Preprints.* Vol. 1, 87–90. London: James & James.

Savarino, P., G. Viscardi, R. Carpignano, E. Barni, and G. Ferrero
1989. Disperse and cationic dyes from aminophenyl-X-azolopyridines. *Dyes and Pigments* 11: 163–72.

Schmelzer, H.
1984. Zusammenhänge zwischen der Wellenlänge der absorbierten Strahlung und der Lichtechtheit von Buntpigmenten (Connection between the wavelength of absorbed radiation and the lightfastness of colored pigments). In *17th Congress of the Fédération d'associations de techniciens des industries des peintures, vernis, émaux et encres d'imprimerie de l'Europe continentale* (FATIPEC), *Lugano, Switzerland, 23–29 September 1984: Preliminary programme*, 499–509. Zurich: Schweizerische Vereinigung der Lack- und Farben.

Schmidt, J. A., E. Goldszmidt, C. Heitner, J. C. Scaiano, A. B. Berinstain, and L. J. Johnston
1993. Photodegradation of α-guaiacoxyacetoveratrone: Triplet-state reactivity induced by protic solvents. In *Photochemistry of Lignocellulosic Materials*, ed. C. Heitner and J. C. Scaiano, 122–28. ACS Symposium Series, no. 531. Washington, D.C.: American Chemical Society.

Schmidt, J. A., C. Heitner, G. P. Kelly, P. A. Leicester, and F. Wilkinson
1991. Diffuse reflectance laser flash photolysis studies of the photochemistry of bleached thermomechanical pulp. *Journal of Photochemistry and Photobiology A: Chemistry* 57: 111–25.

Schmidt, J. A., C. Heitner, G. P. Kelly, and F. Wilkinson
1990. Diffuse reflectance laser flash photolysis of mechanical pulp. Part 1, Detection and identification of transient species in the photolysis of thermomechanical pulp. *Journal of Pulp and Paper Science* 16(4): J111–17.

Schneider, W. E.
1997. Colorimetry: Examining the methods and tools for measuring and analyzing color. In *The Photonics Directory.* Book 3, *The Photonics Design and Applications Handbook, 43rd International Edition*, H52–60. Pittsfield Mass.: Laurin Publishing Co.

Schweizer, R., and B. Mühlethaler
1968. Einige grüne und blaue kupferpigmente. *Farbe + Lack* 74: 1159–73.

Schweppe, H., and H. Roosen-Runge
1986. Carmine. In *Artists' Pigments: A Handbook of Their History and Characterization.* Vol. 1, ed. R. L. Feller, 255–83. Washington, D.C.: National Gallery of Art.

Scopick, D.
1978. *The Gum Bichromate Book*. Rochester, N.Y.: Light Impressions.

Searle, N. D.
1994. Effect of light source emission on durability testing. In *Accelerated and Outdoor Durability Testing of Organic Materials,* ed. W. D. Ketola and D. Grossman, 52–67. Philadelphia: American Society for Testing and Materials.

Selwitz, C.
1988. *Cellulose Nitrate in Conservation*. Marina del Rey, Calif.: Getty Conservation Institute.

Seves, A., T. de Marco, and A. Siciliano
1995. Blending polypropylene with hydrogenated oligocyclopentadiene: A new method for the production of dyeable fibers. *Dyes and Pigments* 28(4): 19–29.

Seybold, G., and G. Wagenblast
1989. New perylene and violanthrone dyestuffs for fluorescent collectors. *Dyes and Pigments* 11: 303–17.

Seymour, R. B., and C. E. Carraher Jr.
1992. *Polymer Chemistry: An Introduction.* 3d ed. New York: Marcel Dekker.

Shah, H., I. B. Rufus, and C. E. Hoyle
1994. Photochemistry of bisphenol-A based polycarbonate: Early detection of photoproducts by fluorescence spectroscopy. *Macromolecules* 27: 553–61.

Shkrob, I. A., M. C. Depew, and J. K. S. Wan
1992. Free radical induced oxidation of alkoxy-phenols: Some insights into the processes of photoyellowing of papers. *Research in Chemical Intermediates* 17: 271–85.

Shore, J.
1990. Dye structure and application properties. Chap. 3 in *Colorants and Auxiliaries: Organic Chemistry and Application Properties.* Vol. 1, *Colorants*, ed. J. Shore, 74–145. Bradford, England: Society of Dyers and Colourists.

Shosenji, H., K. Gotoh, C. Watanabe, and K. Yamada
1983. The effects of fluorescent substances on the photofading of colours. VI–Photochemical interaction between methyl violet and a fluorescent whitening agent of a pyrazoline type in copolymers of acrylonitrile. *Journal of the Society of Dyers and Colourists* 99: 98–101.

Silvy, J., and J. F. Le Nest
1976. A photochromic effect shown during the ageing of a cellulose material by heat and light. In *Fundamental Properties of Paper Related to Its Uses*, ed. F. M. Bolam. Vol. 2, 755–60. London: Ernest Benn.

Sirbiladze, K. J., A. Vig, V. M. Anyisimov, O. M. Anyisimova, G. E. Krichevskiy, and I. Rusznák
1990. Determination of the lightfastness of dyestuffs by kinetic characteristics. *Dyes and Pigments* 14: 23–34.

Siripant, S.
1986. Light stability of colour prints. In *Proceedings of the International Symposium, the Stability and Conservation of Photographic Images: Chemical, Electronic, and Mechanical*, 194–202. Bangkok: Society of Photographic Scientists and Engineers.

Sistach, M. C., and I. Espadaler
1993. Organic and inorganic components of iron gall inks. In *10th Triennial Meeting, Washington, D.C., 3–10 August 1993: ICOM Committee for Conservation Preprints.* Vol. 2, 485–90. Paris: ICOM Committee for Conservation.

Smith, W. F., Jr., W. G. Herkstroeter, and K. L. Eddy
1976. Mechanistic pathways for visible light fading of pyrazolone azomethine dyes in oil-in-gelatin dispersions. *Photographic Science and Engineering* 20: 140–48.

Snell, J. F.
1978. Radiometry and photometry. Section 1 in *Handbook of Optics*, ed. W. G. Driscoll and W. Vaughan, 1–1 to 1–13. New York: McGraw Hill Book Company.

Spikes, J. D.
1977. Photosensitization. Chap. 4 in *The Science of Photobiology*, ed. K. C. Smith, 87–112. New York: Plenum Publishing Co.

Stowe, B. S., V. S. Salvin, R. E. Fornes, and R. D. Gilbert
1973. The effect of near ultraviolet radiation on the morphology of Nylon 66. *Textile Research Journal* 43: 704–14.

Sturge, J., V. Walworth, and A. Shepp, eds.
1989. *Imaging Processes and Materials, Neblette's Eighth Edition.* New York: Van Nostrand Reinhold.

Suppan, P.
1994. *Chemistry and Light.* Cambridge: Royal Society of Chemistry.

Sutherland, L.
1994. North moves south: Conservation solutions for a large collection of Inuit prints and drawings. In *Modern Works, Modern Problems? Conference Papers*, ed. A. Richmond, 122–27. Leigh, Worcester, England: Institute for Paper Conservation.

Tennakone, K., G. R. R. A. Kumara, A. R. Kumarasinghe, P. M. Sirimanne, and K. G. U. Wijayantha
1996. Efficient photosensitization of nanocrystalline TiO_2 films by tannins and related phenolic substances. *Journal of Photochemistry and Photobiology A: Chemistry* 94: 217–20.

Tennent, N. H.
1979. Clear and pigmented epoxy resins for stained glass conservation: Light ageing studies. *Studies in Conservation* 24: 153–64.

Tennent, N. H., and J. H. Townsend
1984. The photofading of dyestuffs in epoxy, polyester and acrylic resins. In *7th Triennial Meeting, Copenhagen, 10–14 September 1984: ICOM Committee for Conservation Preprints,* 84.16.8–16.11. Paris: ICOM Committee for Conservation.

Tennent, N. H., J. H. Townsend, and A. Davis
1982. A simple integrating dosimeter for ultraviolet light. In *Science and Technology in the Service of Conservation: International Institute for Conservation (IIC) Washington Congress Preprints,* 32–38. London: International Institute for Conservation.

Tera, F. M., L. A. Abdou, M. N. Michael, and A. Hebeish
1985a. Fading characteristics of some monoazo dyes on cellulose diacetate and polyamide films, part 1. *Polymer Photochemistry* 6: 361–74.

1985b. Fading characteristics of some monoazo dyes on cellulose diacetate and polyamide films, part 2. *Polymer Photochemistry* 6: 375–83.

Theys, R. D., and G. Sosnovsky
1997. Chemistry and processes of color photography. *Chemical Reviews* 97: 83–132.

Thomas, J. K.
1988. Photophysical and photochemical processes on clay surfaces. *Accounts of Chemical Research* 21: 275–80.

Thomson, G.
1986. Light. Parts 1 and 2 in *The Museum Environment*, 2d ed., 7–10, 164–74. London: Butterworth/Heinemann.

Tokumaru, K., H. Sakuragi, T. Kanno, T. Oguchi, H. Misawa, Y. Shimamura, and Y. Kuriyama

1985. Semiconductor-catalyzed photoreactions of organic compounds. Chap. 3 in *Organic Phototransformations in Nonhomogeneous Media*, ed. M. W. Fox, 43–55. ACS Symposium Series, no. 278. Washington, D.C.: American Chemical Society.

Torikai, A., T. Hattori, and T. Eguchi

1995. Wavelength effect on the photoinduced reaction of polymethyl methacrylate. *Journal of Polymer Science, Polymer Chemistry Edition* 33: 1867–71.

Townsend, J. H., and N. H. Tennent

1993. Colour transparencies: Studies on light fading and storage stability. In *10th Triennial Meeting, Washington, D.C., 3–10 August 1993: ICOM Committee for Conservation Preprints.* Vol. 1, 281–86. Paris: ICOM Committee for Conservation.

Toyoshima, K.

1991. Study of *tegurome-urushi* (handiworked *kurome-urushi*) for restoration (III) (in Japanese; abstract, tables, and figures in English). In *Scientific Papers on Japanese Antiques and Art Crafts*, ed. T. Kitamura, 8–17. Tokyo: Association of Scientific Research on Historic and Artistic Works of Japan.

Trichet, V., S. Grelier, A. Castellan, H. Choudhury, and R. S. Davidson

1996. Attempt to photostabilize paper made from high-yield pulp by application of UV screens in conjunction with thiols. *Journal of Photochemistry and Photobiology A: Chemistry* 95: 181–88.

Trotman, E. R.

1970. Testing dyed materials. Chap. 25 in *Dyeing and Chemical Technology of Textile Fibers*, 4th ed., 587–95. London: Charles Griffin & Co., Ltd.

Turro, N. J., and A. A. Lamola

1977. Photochemistry. Chap. 3 in *The Science of Photobiology*, ed. K. C. Smith, 63–86. New York: Plenum Publishing Co.

Tylli, H., I. Forsskåhl, and C. Olkkonen

1993. A spectroscopic study of photoirradiated cellulose. *Journal of Photochemistry and Photobiology A: Chemistry* 76:143–49.

1995. The effect of photoirradiation on high-yield pulps: Spectroscopy and kinetics. *Journal of Photochemistry and Photobiology A: Chemistry* 87: 181–91.

Umney, N.

1987. Oriental lacquer. *Conservation News* (32): 23–25; (33): 13–15.

van Beek, H. C. A., and P. M. Heertjes

1966. Fading by light of organic dyes on textiles and other materials. *Studies in Conservation* 11: 123–31.

van Oosten, T. B.

1994. Investigation into the degradation of weighted silk. In *Contributions of the Central Research Laboratory to the Field of Conservation and Restoration*, ed. H. Verschoor and J. Mosk, 65–76. Amsterdam: Central Research Laboratory.

van Oosten, T. B., and A. Aten

1996. Life-long guaranteed: The effect of accelerated ageing on Tupperware objects made of polyethylene. In *11th Triennial Meeting, Edinburgh, Scotland, 1–6 September 1996: ICOM Committee for Conservation Preprints.* Vol. 2, 971–77. London: James & James.

Vig, A., I. Rusznák, A. Rockenbauer, L. Koreck, and K. Sirbiladze
1996. Stable free radical formation in photofading processes of reactive dyed cotton fabrics in the presence of finishing agents. *Melliand Textilberichte* 77: 58–59, 62.

Vitale, T.
1998. Light levels used in modern flatbed scanners. *RLG DigiNews* 2 (no. 5). Available online at http://www.rlg.org/preserv.

Waller, R.
1994. Conservation risk assessment: A strategy for managing resources for preventive conservation. In *Preventive Conservation: Practice, Theory and Research: Contributions to the Ottawa Congress Preprints*, 12–16. London: International Institute for Conservation.

Ware, M.
1994. *Mechanisms of Image Deterioration in Early Photographs. The Sensitivity to Light of W. H. F. Talbot's Halide-Fixed Images 1834–1844*. London: Science Museum and National Museum of Photography, Film and Television.

Watanabe, T., T. Takizawa, and K. Honda
1977. Photocatalysis through excitation of adsorbates 1: Highly efficient *N*-deethylation of Rhodamine B adsorbed to CdS. *Journal of Physical Chemistry* 81: 1845–51.

Wayne, R. P.
1988. *Principles and Applications of Photochemistry*. Oxford: Oxford Univ. Press.

Webb, M.
1997. Light degradation of Oriental lacquerware—or, Has your lacquer lost its lustre? Paper presented at the 15th Annual Meeting of the American Institute for Conservation of Historic and Artistic Works (AIC), Vancouver, B.C.

2000. Degradation. Chap. 6 in *Lacquer: Technology and Conservation: A Comprehensive Guide to the Technology and Conservation of Asian and European Lacquer*. Oxford: Butterworth/Heinemann.

White, H. S.
1987. Book copiers: Past, present and future. *Restaurator* 8: 18–28.

Whitmore, P. M., and C. Bailie
1990. Studies on the photochemical stability of synthetic resin-based retouching paints: The effects of white pigments and extenders. In *Cleaning, Retouching and Coatings: Technology and Practice for Easel Paintings and Polychrome Sculpture*, ed. J. S. Mills and P. Smith, 144–49. London: International Institute for Conservation.

1997. Further studies on transparent glaze fading: Chemical and appearance kinetics. *Journal of the American Institute for Conservation* 36: 207–30.

Whitmore, P. M., and J. Bogaard
1994. Determination of the cellulose scission route in the hydrolytic and oxidative degradation of paper. *Restaurator* 15: 26–45.

Whitmore, P. M., V. G. Colaluca, and E. Farrell
1996. A note on the origin of turbidity in films of an artists' acrylic paint medium. *Studies in Conservation* 41: 250–55.

Whitmore, P. M., X. Pan, and C. Bailie
1999. Predicting the fading of objects: Identification of fugitive colorants through

direct nondestructive lightfastness measurements. *Journal of the American Institute for Conservation* 38: 395–409.

Wilhelm, H.
1981. Monitoring the fading and staining of color photographic prints. *Journal of the American Institute for Conservation* 21: 49–64.

1993. *The Permanence and Care of Color Photographs: Traditional and Digital Color Prints, Color Negatives, Slides, and Motion Pictures.* Grinnell, Iowa: Preservation Publishing Co.

Wilkinson, F.
1986. Diffuse reflectance flash photolysis. *Journal of the Chemical Society, Faraday Transactions 2: Molecular and Chemical Physics* 82: 2073–81.

Wilkinson, F., A. Goodwin, and D. R. Worrall
1993. Diffuse reflectance laser flash photolysis of thermomechanical pulp. In *Photochemistry of Lignocellulosic Materials*, ed. C. Heitner and J. C. Scaiano, 86–98. ACS Symposium Series, no. 531. Washington, D.C.: American Chemical Society.

Wilkinson, F., and G. P. Kelly
1989. Laser flash photolysis on solid surfaces. Sec. 2.2 in *Photochemistry on Solid Surfaces*, ed. M. Anpo and T. Matsuura, 30–47. Amsterdam: Elsevier.

Wilkinson, F., D. J. McGarvey, and D. R. Worrall
1992. The application of diffuse reflectance laser photolysis to study photochemistry at interfaces and in dyed fabrics. *Proceedings of the Indian Academy of Sciences (Chemical Sciences)* 104: 739–45.

Wilkinson, F., and C. J. Willsher
1987. The application of diffuse reflectance laser flash photolysis to metal phthalocyanines in an opaque environment. In *Photochemistry and Photophysics of Coordination Compounds*, ed. H. Yersin and A. Volger, 327–30. Berlin: Springer Verlag.

Wilkinson, F., C. J. Willsher, and R. B. Pritchard
1985. Laser flash photolysis of dyed fabrics and polymers—I: Rose bengal as a photosensitizing dye. *European Polymer Journal* 21: 333–41.

Wilkinson, F., D. Worrall, D. McGarvey, A. Goodwin, and A. Langley
1993. Kinetic spectroscopy of pyrazolotriazole azomethine dyes. *Journal of the Chemical Society, Faraday Transactions* 89: 2385–90.

Williams, D. R.
1994. Data conversion: A tutorial in electronic document imaging. Tutorial C in *Digital Imaging Technology for Preservation: Proceedings from an RLG Symposium, Held March 17 and 18, 1994, Cornell University, Ithaca, New York,* ed. N. E. Elkington. Mountain View, Calif.: Research Libraries Group.

Wouters, J.
1993. High-performance liquid chromatography of vegetable tannins extracted from new and old leather. In *10th Triennial Meeting, Washington, D.C., 22–27 August 1993: ICOM Committee for Conservation Preprints.* Vol. 2, 669–73. Paris: ICOM Committee for Conservation.

Yagami, K., M. Nakagawa, H. Shosenji, and K. Yamada
1988. The effects of fluorescent substances on the photofading of colours. Part 7: Photofading interaction between azo dyes and fluorescent brightening agents of a

stilbene type on cotton and wool and effect of the treatment with a nickel salt (in Japanese; English abstract). *Sen-i Gakkaishi (Journal of the Society of Fiber Science and Technology of Japan)* 44: 27–31.

Yamada, K., H. Shosenji, and K. Gotoh
1977. The effects of fluorescent substances on the photofading of colours II—The effects of fluorescent brightening agents on the photofading of triphenylmethane dyes. *Journal of the Society of Dyers and Colourists* 93: 219–23.

Yamada, K., H. Shosenji, Y. Nakano, M. Uemura, S. Uto, and M. Fukushima
1981. The effects of fluorescent substances of the photofading of colours. Part 4—Photochemical interaction between fluorescent brightening agent of a pyrazoline type and 1,4-diaminoanthraquinone. *Dyes and Pigments* 2: 21–29.

Zamotaev, P., O. Mityukhin, and S. Luzgarev
1992. 9,10-anthraquinone derivatives as photoinitiators of crosslinking or degradation of polyethylene. *Polymer Degradation and Stability* 35: 195–210.

Zinn, E., J. M. Reilly, P. Z. Adelstein, and D. W. Nishimura
1994. Air pollution effects on library microforms. In *Preventive Conservation: Practice, Theory and Research*, ed. A. Roy and P. Smith, 195–201. London: International Institute for Conservation.

Zollinger, H.
1987a. Colorants for imaging and data recording systems. Chap. 14 in *Color Chemistry: Synthesis, Properties and Applications of Organic Dyes and Pigments*, 283–303. Weinheim, Germany: VCH Publishers, Inc.

1987b. Photo-, thermo- and electrochemical reactions of colorants. Chap. 13 in *Color Chemistry: Synthesis, Properties and Applications of Organic Dyes and Pigments*, 245–81. Weinheim, Germany: VCH Publishers, Inc.

Zweig, A., and W. A. Henderson Jr.
1975. Singlet oxygen and polymer photooxidations I: Sensitizers, quenchers, and reactants. *Journal of Polymer Science, Polymer Chemistry Edition* 13: 717–36.

Index

Note: Page numbers followed by f *and* t *denote figures and tables, respectively.*

About the Author

Terry Trosper Schaeffer received her Ph.D. in biophysics from the University of California, Berkeley, and spent more than two decades doing research on photosynthetic pigments and ion transport across mammalian cell membranes. In 1989 Schaeffer entered the field of conservation research as a Fellow-by-Courtesy at the G. W. C. Whiting School of Engineering at Johns Hopkins University and a visiting research collaborator at the Conservation Analytical Laboratory, Smithsonian Institution (now the Smithsonian Center for Materials Research and Education). She has been a special conservation research fellow in the Conservation Center of the Los Angeles County Museum of Art (LACMA) and an outside contractor with the Getty Conservation Institute and is currently the chemical hygiene officer in the Conservation Center at LACMA.